W9-BJC-983

THE

LEAFCUTTER
ANTS

BOOKS BY THE AUTHORS

ALSO BY BERT HÖLLDOBLER AND EDWARD O. WILSON

The Ants (1990); Pulitzer Prize, General Nonfiction, 1991
Journey to the Ants: A Story of Scientific Exploration (1994)
*The Superorganism: The Beauty, Elegance, and Strangeness
   of Insect Societies* (2009)

ALSO BY BERT HÖLLDOBLER

Experimental Behavioral Ecology and Sociobiology, with Martin Lindauer, edi-
   tors (1985)
Herbivory of Leaf-Cutting Ants: A Case Study on Atta colombica *in the Tropical
   Rainforest of Panama,* with Rainer Wirth, Hubert Herz, Ronald
   J. Ryel, and Wolfram Beyschlag (2003)

ALSO BY EDWARD O. WILSON

The Theory of Island Biogeography, with Robert H. MacArthur (1967); new preface, 2001

A Primer of Population Biology, with William H. Bossert (1971)

The Insect Societies (1971)

Sociobiology: The New Synthesis (1975); new edition, 2000

On Human Nature (1978); Pulitzer Prize, General Nonfiction, 1979

Caste and Ecology in the Social Insects, with George F. Oster (1978)

Genes, Mind, and Culture, with Charles J. Lumsden (1981)

Promethean Fire: Reflections on the Origin of Mind, with Charles J. Lumsden (1983)

Biophilia (1984)

Success and Dominance in Ecosystems: The Case of the Social Insects (1990)

The Diversity of Life (1992)

Naturalist (1994); new edition, 2006

In Search of Nature (1996)

Consilience: The Unity of Knowledge (1998)

Biological Diversity: The Oldest Human Heritage (1999)

The Future of Life (2002)

Pheidole in the New World: A Hyperdiverse Ant Genus (2003)

From So Simple a Beginning: The Four Great Books of Darwin, edited with introductions (2005)

Nature Revealed: Selected Writings, 1949–2006 (2006)

The Creation: An Appeal to Save Life on Earth (2006)

Anthill: A Novel (2010)

THE
LEAFCUTTER ANTS

Civilization by Instinct

BERT HÖLLDOBLER

AND

EDWARD O. WILSON

W. W. NORTON & COMPANY

NEW YORK · LONDON

Frontispiece: Diversity of plant fragments collected by one *Atta colombica* colony in Panama during one year. Artistic arrangement and photograph by Christian Ziegler. (© Christian Ziegler / DanitaDelimont.com.)

Copyright © 2011 by Bert Hölldobler and Edward O. Wilson

Parts of this text have previously appeared in Chapter 9 of *The Superorganism* by Bert Hölldobler and Edward O. Wilson

All rights reserved
Printed in China
First Edition

For information about permission to reproduce selections from this book, write to Permissions, W. W. Norton & Company, Inc., 500 Fifth Avenue, New York, NY 10110

For information about special discounts for bulk purchases, please contact W. W. Norton Special Sales at specialsales@wwnorton.com or 800-233-4830

Manufacturing by RR Donnelley, Shenzhen
Book design by Abbate Design
Production manager: Devon Zahn

Library of Congress Cataloging-in-Publication Data

Hölldobler, Bert, 1936–
The leafcutter ants : civilization by instinct / Bert Hölldobler and Edward O. Wilson. — 1st ed.
p. cm.
Includes bibliographical references and index.
ISBN 978-0-393-33868-3 (pbk.)
1. Leaf-cutting ants. I. Wilson, Edward O. II. Title.
QL568.F7H577 2011
595.79'6—dc22

2010016202

W. W. Norton & Company, Inc.
500 Fifth Avenue, New York, N.Y. 10110
www.wwnorton.com

W. W. Norton & Company Ltd.
Castle House, 75/76 Wells Street, London W1T 3QT

1 2 3 4 5 6 7 8 9 0

CONTENTS

LIST OF ILLUSTRATIONS

FIGURES

PLATES

THE
LEAFCUTTER
ANTS

PROLOGUE

—————————

If a congress of naturalists were to gather to choose the seven wonders of the animal world, they would be compelled to include the bizarre and mighty civilizations of the attine leafcutters. Throughout the tropical and subtropical regions of the New World, these insects dominate the forests, grasslands, and pastures.

Anywhere you travel on the mainlands of Central and South America, from the wild interior to plazas and vacant lots in the hearts of the cities, you will soon encounter leafcutters. What catches your attention at first are massive lines of relatively large reddish brown worker ants. They run in columns as wide as ten ants abreast, and as tight as soldiers double-timing in a parade. They travel on miniature highways the width of a human hand, which they keep bare of vegetation and debris. Some are outward bound; a roughly equal number are homeward bound. Most among the latter carry a freshly cut section of a leaf or flower petal, which they grip in their mandibles and direct back over their bodies like umbrellas. These are the "parasol ants," local people in Texas and Louisiana will tell you.

Look closely at the burdened ants, and you are likely to see pygmy replicas riding as hitchhikers on the transported leaf fragments. Do these miniature ants act like mahouts to guide their big nestmates home? No, their role is even stranger; they serve as living flywhisks. Ant columns attract parasitic flies that are deadly. They descend like dive-bombers and, if unimpeded, lay eggs upon or near the necks of the big ants. The maggots that soon hatch work their way into the bodies of the ants and consume their tissues. The flywhisk guardlings prevent this action by standing on top of their leaf-carrying sisters and striking at the flies with their front legs and swatting mandibles.

If you follow the caravan of laden ants, they will bring you to the nest. It might be fifty or even more than a hundred meters down the trail. The trip may take you through dense undergrowth, and perhaps a steep little ravine or two. Inevitably and often suddenly, the nest comes into view. It is a city of millions, a subterranean metropolis. Above it is a circular dome of excavated soil that rises two meters or more. In the underground, the ants have excavated thousands of chambers, roughly the size of a human head, more precisely from a thirtieth of a liter to fifty liters in volume. All are connected by a labyrinth of tunnels. The chambers are filled with a fluffy gray mass. Their thin walls are built to create a maximum of surface per unit of volume. Upon the walls grows a species of fungus that exists exclusively in symbiosis with agricultural ants such as the leafcutters and their evolutionarily less advanced close relatives. The fungus only rarely develops a mushroom with a stalk and cap. Instead, it proliferates in the form of mats of threadlike hyphae.

The walls of the cells upon which the fungus feeds are built from a paste, like papier-mâché. This substance is manufactured from the fragments of vegetation brought in by the foraging workers.

Aside from plant sap obtained from freshly cut vegetation, the leafcutters subsist entirely on the cultivated fungus. They have invented a method of converting fresh vegetation, a material the digestive system of ants cannot handle, into an edible food product. The leafcutter ants are partly comparable in achievement to that of human agriculturists. And they have attained a break-through of organic evolution: by utilizing fresh vegetation on which to grow their crops, they have tapped into a virtually unlimited food source.

The interdependence of the leafcutters and their fungus is one of the most successful symbioses of all time. The ants are the principal consumers of living plant material and therefore a dominant force in the land environments of the American tropics. A full-grown colony consumes approximately the same quantity of plant material as a cow. Through much of tropical America and wherever the ants can invade gardens and cropland, they are the principal insect pests of agriculture.

And so all within their great range know them. They are the *saúva* of Brazil, *isaú* of Paraguay, *cushi* of Guyana, *zampopo* of Costa Rica, *wee-wee* of Nicaragua and Belize, *cuatalata* of Mexico, *bibijagua* of Cuba, and town ant or parasol ant of Texas and Louisiana.

Leafcutter colonies can best be understood scientifically as complex organic structures with a single purpose: the conversion of plant life into more colonies of leafcutter ants. They are civilizations designed by natural selection to replicate themselves in as many copies as possible before their inevitable death. Because they possess one of the most complex communication systems known in animals, as well as the most elaborate caste systems, air-conditioned nest architecture, and populations into the millions, they deserve recognition as Earth's ultimate superorganisms. They are all the more remarkable for consisting entirely of a mother queen and her daughters. Males are reared only seasonally, are then in a tiny minority, and serve merely to inseminate virgin queens during nuptial flights away from the nests. Then they die, by design of their bodies and instinctive behavior.

If visitors from another star system had visited Earth a million years ago, before the rise of humanity, they might have concluded that leafcutter colonies were the most advanced societies this planet would ever be able to produce. Yet there was one step to take, the invention of culture, making it possible to write this book about them.

1

THE ULTIMATE
SUPERORGANISMS

No one can give us an exact number of animal species living on Earth today, but all biologists agree that millions more species exist than the approximately 1.9 million that have been described so far. Quantitative faunistic investigations in many habitats suggest about 8 million extant species; other assessments claim 30 million species or even more.[1] Most of these species that share mother earth with us are still unknown to science and, sadly, may never become known because of ongoing man-made habitat destruction and ensuing species extinction.

About half of all described animal species are insects, at approximately 900,000 species. Of this assemblage only about 2 percent live in the most advanced, or "eusocial," systems. We consider an insect society to be eusocial when the following criteria are met: cooperative care for the immature individuals, overlap of at least two generations in the same society, and the coexistence of reproductive and nonreproductive members.

In every species at the evolutionarily advanced grade of eusociality, we find distinct morphological or at least physiological castes: a reproductive caste (queen) and a nonreproductive caste (worker). The latter can be further subdivided into several morphologically distinct subcastes, such as minor, media, and major workers. In advanced eusocial societies, workers exhibit a sophisticated system of division of labor, whereby task and role assignments can depend either on the age of the individual ("age polyethism") or on morphological features ("physical polyethism"), or on both. Depending on the species-specific social organization and stage of development, the colony (society) in the distribution of roles and tasks among the workers is more or less flexible. Nevertheless, a good rule of thumb is that young individuals spend most of their time inside the nest, being mainly engaged in the care of the queen and brood, whereas older workers are more involved with risky activities outside the nest, such as midden work, nest construction, foraging, and defense of the nest and territory.

The combination of cooperation and division of labor bestows a tremendous advantage upon the social insects. Where at any given moment a solitary organism can be doing only a few things and can be in just one place, an insect society can, by deploying its worker force, perform many activities and can be in several different places simultaneously. As a consequence, social insects, especially ants and termites, play a dominant role in most land ecosystems. Although only 2 percent of the known insect species are eusocial, they are the principal predators and scavengers of small animals, turners of soil, and prey for other animals.

For example, Ernst Josef Fittkau and H. Klinge[2] found that in the Brazilian terra firme forest ants and termites (all species of which live in societies) together compose roughly 30 percent of the entire animal biomass. When stingless bees and polybiine wasps are added, all the social insects together make up more than 75 percent of the entire insect biomass. On the basis of data, and other evidence from natural history, we suggest that these organisms, and particularly the ants and termites, occupy center stage in the terrestrial environment and have done so for tens of millions of years around the world. They have pushed out solitary insects from the generally most favorable nest sites. The solitary forms occupy the more distant twigs, the very moist or dry excessively crumbling pieces of wood, the surface of leaves—in short, the more remote and transient nesting places. At the risk of oversimplification, the picture we see is the following: social insects hold the ecological center; solitary insects occupy the periphery.

The ants are divided at the present time into nineteen taxonomic subfamilies. They exhibit by far the most impressive adaptive diversity of any of the eusocial insect groups. Close to 14,000 species are known to science, but on the basis of the rate of discovery of species, systematists estimate that as many as 25,000 exist. Although all ant species are eusocial, the social organizations of particular species groups vary greatly. In some species, for example, each colony has only one queen ("monogynous"), while in others each colony has many queens ("polygynous"). Some species form colonies that consist of a relatively small number of workers (50–200), while other species form colonies containing hundreds of thousands or even, as in some of the leafcutters, millions of workers. Equally diverse are the modes of colony foundation and colony reproduction as well as the ways in which the division of labor systems are organized, how communication among the individuals and groups of individuals functions, and, finally, how the colonies forage and which resources they exploit.[3]

Myrmecologists—scientists who study ants—recognize several pinnacles in ant evolution, including the army ants of the Neotropics and the driver ants of Africa; the tree-dwelling weaver ants of the genus *Oecophylla* of Africa, Asia, and Australia; and the "supercolonies" of the ant *Formica yessensis* (one of which on the Ishikari Coast of Hokkaido was found to comprise 306 million workers and 1,080,000 queens living in 45,000 interconnected nests across a territory of 2.7 square kilometers).[4] Also to be included are the migrating herdsmen of the genus *Dolichoderus* (= *Hypoclinea*) in the rainforest of the Malaysian peninsula[5] and, far from least, the fungus growers of the myrmicine tribe Attini. This is by no means a complete listing of the astounding diversity of lifestyles in the ants; it is merely a sampling of the most spectacular ways of life in which the ants have evolved over the course of their approximately 120 million years of evolutionary history.

During this evolution, the brain capacity of individual ants has probably been pushed close to the limit. Beyond a certain limit, the evolving lines built advances upon social organization. The amazing feats of army ants, weaver ants, or leafcutter ants come not from complex behavior of individual colony members but from the concerted actions of many nestmates working together. To watch a single ant wandering away from her colony is to see at most a huntress in the field or a small creature of ordinary demeanor digging a hole in the ground. By itself this one ant is, however, a vast disappointment. It is really no ant at all. What counts is the entire colony, which is the equivalent of an entire nonsocial organism. The colony is the unit that we must examine in order to understand the biology and evolution of both the colony and the ant that is part of it.

It is not just analogy and metaphor to speak of a superorganism, and therefore to invite a detailed comparison between the society and the conventional organism. The concept of a level above the organism was extremely popular in the early part of the twentieth century. The great American entomologist William Morton Wheeler, like many of his contemporaries, returned to it repeatedly in his writings. In his influential 1911 essay, "The Ant Colony as an Organism," he stated that the ant colony is really an organism and not merely the analog of one. It behaves, he said, as a unit. It possesses distinctive properties of size, behavior, and organization that are transmitted from one generation to the next. The queen is the reproductive organ; the workers are the supporting brain, heart, gut, and other tissues. The exchange of liquid and food among colony members is the equivalent of the circulation of blood and lymph. This exercise, however elaborate or inspirational, eventually exhausted its possibilities. The limitations of the approach based primarily on analogy became increasingly obvious as biologists discovered more of the fine details of communication and caste formation and division of labor that lie at the heart of colonial organization. By 1960, the expression "superorganism" had all but vanished from the vocabulary of the scientists.

Old ideas in science, however, seldom die. Like the mythical Greek giant Antaeus, many instead merely fall to Earth, where they gain new strength and rise again. With a far greater knowledge of both organisms and colonies available now, there has been increasing acceptance of the view that the entire colony represents an extended phenotype of the queen and her mates on which evolutionary selection operates. Comparisons of these two levels of biological organization, organism and superorganism, have been resumed with greater depth and precision.[6] The new exercise has a goal larger than just the intellectual delectations of analogy. It now allows us to mesh information from developmental biology with that from the study of animal societies in order to uncover general and exact principles of biological organization. The question of general interest for biology is the similarities, the joint rules and algorithms, arising between morphogenesis and sociogenesis.

Ant colonies are more than the sum of their parts. They are

operational units with emergent traits that arise from complex interactions of colony members. The ultimate possibilities of superorganismic evolution are perhaps best expressed by the spectacular leafcutter ants of the genera *Atta* and *Acromyrmex,* which we will now lay before you.

THE ATTINE
BREAKTHROUGH

Both human civilization and the evolution of extreme insect superorganisms were attained by agriculture, a form of mutualistic symbiosis of animals with plants or fungi. Human agriculture, which originated about 10,000 years ago, was a major cultural transition that catapulted our species from a hunter-gatherer lifestyle to a technological and increasingly urban existence, accompanied by an enormous expansion of population. Humanity thereby turned itself into a geophysical force and began to alter the environment of the entire planetary surface.

Approximately 50 to 60 million years before this momentous shift, some social insects had already made the evolutionary transition from a hunter-gatherer existence to agriculture. In particular, macrotermitine termites in the Old World and attine ants in the New World invented the culturing of fungi, which then became an essential part of their diet.

Many of the fungus-growing termites of the subfamily Macrotermitinae are concentrated in tropical Africa and Southeast Asia. They construct tremendous nests that have been called castles of clay (Plate 1). The excrement of these insects is built into fungus combs, which occupy

PLATE 1. Closely spaced large nest mounds of the fungus-growing termite *Macrotermes bellicosus* in Gomoe National Park, Ivory Coast, West Africa. (Photo: Manfred Kaib.)

special gardens in the center of the nest (Plate 2). The combs resemble sponges, with numerous convoluted ridges and tunnels. The basidiomycete fungus sprouts round white spherules out from the substratum, but these are not especially favored as food bodies. Instead, the termites consume everything in the comb: substratum, mycelia, spherules, and all. In a special royal chamber built from cementlike material reside, well protected, the gigantic queen and her "king" surrounded by different stages of workers and soldiers (Plate 3). In contrast to ant societies and those of other social hymenopterans (such as bees and wasps) that are female societies (males live only to inseminate the reproductive females and then die), termite societies consist of females and males.

PLATE 2. *From upper left, clockwise*: Nest of *Macrotermes michaelsenki* broken open to expose the fungus garden in the center of the nest. Fungus combs in the fungus garden of a *Macrotermes michaelsenki* nest. The white dots on the combs are the fungal spherules. Close-up picture of the fungus combs and spherules in the nest of *Macrotermes michaelsenki*. (Photos: Manfred Kaib.)

PLATE 3. Queen of *Macrotermes michaelsenki* surrounded by workers, soldiers, and worker larvae inside the "royal chamber." (Photo: Manfred Kaib.)

These most advanced agricultural insect societies, like their human counterparts, rose to ecological dominance. The trend is especially marked in the leafcutter ants.[7]

Whereas most fungus-growing attine ants, composing the anatomically and behaviorally "primitive" species, gather and process rotting leaf fragments and dead organic material on which they grow their specific fungi, the evolutionary invention of cutting and harvesting live plant material opened up a huge new nutritional niche for the species of *Acromyrmex* and *Atta*. As in human history, the innovation propelled further evolutionary development.[8]

The tribe Attini is a morphologically very distinctive group limited to the New World. Most of its thirteen genera and approximately 230 species occur in tropical regions of Mexico and Central and South America. Some species occur in the southern portions of the United States, and several are adapted to arid habitats in the southwestern states. One species, *Trachymyrmex septentrionalis,* ranges north to the pine barrens of New Jersey, while in the opposite direction, several species of *Acromyrmex* penetrate to the cold-temperature deserts of central Argentina.[9]

The attine ants, all of which are fungus growers, are overall a monophyletic group—that is, they originated from one common ancestor species. The two genera of the so-called leafcutter or leaf-cutting ants, *Acromyrmex* and *Atta,* are combined with three additional genera in the derived, monophyletic group of the "higher attines." The remaining eight genera are assembled in the "lower attines," a paraphyletic assemblage of basal lineages.[10]

Most cultivated fungi belong to the basidiomycete family Lepiotaceae (Agaricales: Basidiomycota), with the great majority belonging to two genera, *Leucoagaricus* and *Leucocoprinus* (Leucocoprineae).[11] Ulrich Mueller and his co-workers reason that because "most basal attine lineages cultivate leucocoprineous mutualists, attine fungi culture likely originated with the cultivation of leucocoprineous fungi."[12] The majority of fungi of the lower attines are a polyphyletic mix within the family Lepiotaceae, representing two clades.[13] There are a couple of remarkable exceptions to this fidelity: some species of the attine genus *Apterostigma* have secondarily changed to nonlepiotaceous fungi that belong to the family Tricholomataceae.[14] In addition, a small group of lower attine ants cultivate yeast, and in a unicellular phase. Contrary to previous assumptions, this is not an ancestral state of fungus growing. A cladistic analysis of yeast-culturing attine ants (*Cyphomyrmex rimosus* group) (Plate 4, upper) revealed that this clade is not basal but actually derived within the lower attine ants.[15] Finally, all higher attine ants cultivate a derived monophyletic group of fungi in the tribe Leucocoprineae.

To summarize: on the basis of particular morphological features of attine ants (all fungus-growing ants belong to the tribe Attini), their exclusive occurrence in the New World, and their abundance in the Neotropics, researchers have proposed that ant agriculture originated a single time in South America early in the period after its isolation from Africa. In a recent comprehensive analysis, Ted Schultz and Sean Brady produced a time-calibrated phylogeny of attine fungus-growing ants with age estimates for the origins of the ant agricultural systems. This made it possible to reconstruct the major evolutionary transitions in fungus-growing ants that culminated in the emergence of the leafcutter ants of the genera *Acromyrmex* and *Atta*. Five distinct agricultural systems in attine ants were identified. The original mode of fungus cultivation involves a diverse subset of fungus species that all belong to the tribe Leucocoprineae ("parasol mushrooms") and that evolved approximately 60 to 50 million years ago in the leaf litter stratum. As we already mentioned, the original fungus-growing ants were not leafcutters, but collected vegetable debris, small wilted plant remains that the ants retrieved to their nest and from which they prepared a compost to cultivate diverse leucocoprineous fungi species. These cultivated fungi "are identical or closely related to free living fungal populations."[16] Colonies of these basal attine species are small, consisting of a few hundred worker ants and usually one queen. This "lower agriculture" mode can be found in extant species of the genera *Apterostigma* (Plate 4, lower), *Cyphomyrmex* (Plate 5) *Mycocepurus* (Plate 6), *Mycetosoritis* (Plate 7), *Mycetarotes* (Plate 8), and several more (see Plate 9).

PLATE 4 (*facing page*). *Upper*: Queen of the fungus-growing ant *Cyphomyrmex rimosus*, collected in southern Florida. *Lower*: The fungus-growing ants *Apterostigma auriculatum* from Panama cultivate their fungus with organic detritus. (Photos: Alex Wild.)

PLATE 5. *Upper*: *Cyphomyrmex wheeleri* workers tend their fungus garden. This species has a wide distribution in temperate regions and can be found in California far north of San Francisco. *Lower*: Winged queens and males in the fungus garden of *Cyphomyrmex wheeleri*. The alates will soon leave the nest for the mating flight. (Photos: Alex Wild.)

PLATE 6. *Upper*: Worker of the fungus-growing ant *Mycocepurus curvispinosus* from Panama. *Lower*: Inside the fungus garden of *Mycocepurus smithi* from Panama. (Photos: Alex Wild.)

PLATE 7. *Upper*: Worker of the fungus-growing ant *Mycetosoritis hartmanni* from Texas. *Lower*: Fungus garden of *Mycetosoritis hartmanni*. The ants feed the fungus with organic debris collected outside the nest. (Photos: Alex Wild.)

PLATE 8. *Upper*: Foraging worker of *Mycetarotes* sp. from the National Park Iguazu in northern Argentina. *Lower*: A *Mycetarotes* male. Note the large eyes and long antennae typical for males in many ant species. Most likely these large sensory organs aid males in locating females during the mating flight. (Photos: Alex Wild.)

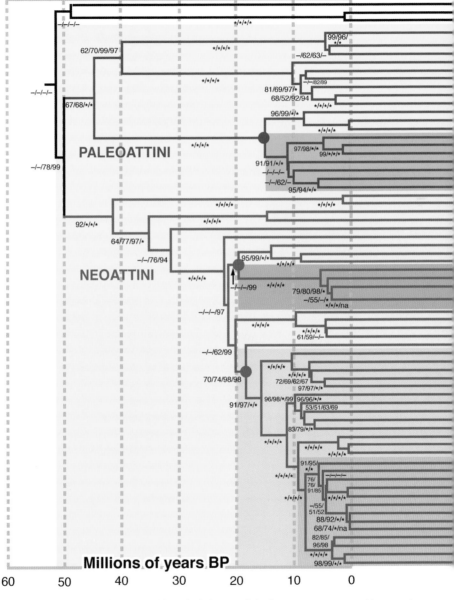

PLATE 9. *Two-page spread*: A time-calibrated phylogeny of the fungus-growing ants with age estimates for the origins of the five known ant agricultural systems. Agricultural systems, indicated by colored rectangles, are defined by phylogenetically distinct groups of associated fungal cultivars. (Illustration: Ted Schultz, modified from T. R. Schultz and S. G. Brady, "Major evolutionary transitions in ant agriculture," *Proceedings of the National Academy of Sciences USA* 105[14]: 5435–5440 [2008].)

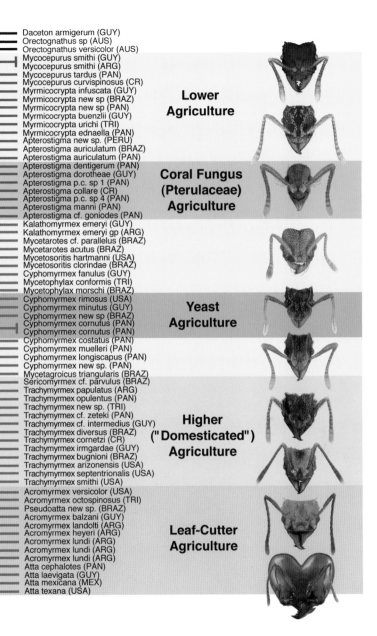

Daceton armigerum (GUY)
Orectognathus sp (AUS)
Orectognathus versicolor (AUS)
Mycocepurus smithi (GUY)
Mycocepurus smithi (ARG)
Mycocepurus tardus (PAN)
Mycocepurus curvispinosus (CR)
Myrmicocrypta infuscata (GUY)
Myrmicocrypta new sp (BRAZ)
Myrmicocrypta new sp (PAN)
Myrmicocrypta buenzlii (GUY)
Myrmicocrypta urichi (TRI)
Myrmicocrypta ednaella (PAN)
Apterostigma new sp. (PERU)
Apterostigma auriculatum (BRAZ)
Apterostigma auriculatum (PAN)
Apterostigma dentigerum (PAN)
Apterostigma dorotheae (GUY)
Apterostigma p.c. sp 1 (PAN)
Apterostigma collare (CR)
Apterostigma p.c. sp 4 (PAN)
Apterostigma manni (PAN)
Apterostigma cf. goniodes (PAN)
Kalathomyrmex emeryi (GUY)
Kalathomyrmex emeryi gp (ARG)
Mycetarotes cf. parallelus (BRAZ)
Mycetarotes acutus (BRAZ)
Mycetosoritis hartmanni (USA)
Mycetosoritis clorindae (BRAZ)
Cyphomyrmex fanulus (GUY)
Mycetophylax conformis (TRI)
Mycetophylax morschi (BRAZ)
Cyphomyrmex rimosus (USA)
Cyphomyrmex minutus (GUY)
Cyphomyrmex new sp (BRAZ)
Cyphomyrmex cornutus (PAN)
Cyphomyrmex cornutus (PAN)
Cyphomyrmex costatus (PAN)
Cyphomyrmex muelleri (PAN)
Cyphomyrmex longiscapus (PAN)
Cyphomyrmex new sp. (PAN)
Mycetagroicus triangularis (BRAZ)
Sericomyrmex cf. parvulus (BRAZ)
Trachymyrmex papulatus (ARG)
Trachymyrmex opulentus (PAN)
Trachymyrmex new sp. (TRI)
Trachymyrmex cf. zeteki (PAN)
Trachymyrmex cf. intermedius (GUY)
Trachymyrmex diversus (BRAZ)
Trachymyrmex cornetzi (CR)
Trachymyrmex irmgardae (GUY)
Trachymyrmex bugnioni (BRAZ)
Trachymyrmex arizonensis (USA)
Trachymyrmex septentrionalis (USA)
Trachymyrmex smithi (USA)
Acromyrmex versicolor (USA)
Acromyrmex octospinosus (TRI)
Pseudoatta new sp. (BRAZ)
Acromyrmex balzani (GUY)
Acromyrmex landolti (ARG)
Acromyrmex heyeri (ARG)
Acromyrmex lundi (ARG)
Acromyrmex lundi (ARG)
Acromyrmex lundi (ARG)
Atta cephalotes (PAN)
Atta laevigata (GUY)
Atta mexicana (MEX)
Atta texana (USA)

Lower Agriculture

Coral Fungus (Pterulaceae) Agriculture

Yeast Agriculture

Higher ("Domesticated") Agriculture

Leaf-Cutter Agriculture

PLATE 10. *Upper*: Worker of *Sericomyrmex amabilis* in the fungus garden. The ants were collected in Panama. *Lower*: Fungus garden of *Trachymyrmex desertorum* from southern Arizona. (Photos: Alex Wild.)

PLATE 11. *Upper*: Foundress queen of *Trachymyrmex arizonensis* in her incipient fungus garden, collected near Tucson, Arizona. *Lower*: A forager of *Trachymyrmex carinatus* carries a flower petal to the nest, where it will be fed to the fungus. The picture was taken in the field near Portal, Arizona. (Photos: Alex Wild.)

PLATE 12. Workers of the leafcutter ant *Acromyrmex striatus* from Argentina. (Photo: Alex Wild.)

From this ancestral fungi culture of attine ants arose three fungi cultural systems during a period of 30 million years, each comprising phylogenetic distinct fungal cultivars. From one of those derived systems (the "higher 'domesticated' agriculture" of the *Trachymyrmex* group) evolved the leafcutter agriculture that employs a single fungal species.

The *Trachymyrmex* group (which includes *Sericomyrmex*) (Plates 10 and 11) is a phylogenetic sister group to the leafcutter ants *Acromyrmex* (Plate 12) and *Atta* (Plates 13 to 15)—that is, they split off from a common ancestor species about 17 million years ago. In particular, the *Trachymyrmex septentrionalis* species group is closely related to the leafcutter ant genera, *Acromyrmex*

PLATE 13 (*facing page*). Two *Atta sexdens* media workers cooperate to cut a piece from a live twig of a plant. (Photo: Bert Hölldobler.)

and *Atta,* and is considered their closest sister group, both diverging from a common ancestor species about 8–12 million years ago. This divergence marks one of the most significant evolutionary transitions ("breakthroughs" is perhaps the better term) from "generalized higher agriculture" to leaf- or grass-cutting agriculture. In fact, Schultz and Brady report observations of occasional leaf cutting in species of this *Trachymyrmex* group.

It has long been assumed that the fungal transmission is strictly vertical, that is, a transfer of fungal cultivars from parent nests to offspring nests. This would imply that the clonally propagated fungal lineages evolved in parallel with the lineages of the ant mutualists over millions of years. However, at least some of the lower attines propagate cultivars that were recently domesticated from free-living populations of Lepiotaceae.[17] Although the "higher attines" are still thought to propagate ancient clones several million years old,[18] how ancient these clones really are remains uncertain. In fact, patterns of lateral transfer of fungal cultivars have been demonstrated in species of the lower attine genus *Cyphomyrmex* (see Plate 5). Laboratory colonies deprived of their fungus garden regained cultivars either by joining a neighboring colony, by stealing a neighbor's garden, or by invading such a garden. As we will show later, pathogens can devastate gardens of attine colonies under natural conditions. Joining, stealing, or usurping neighbors' gardens is probably an important adaptation to counter garden loss, which otherwise would be a fatal catastrophe for the afflicted colony.[19] Similar patterns of occasional lateral transfer of fungus material have been demonstrated between two sympatric *Acromyrmex* species.[20]

PLATE 14 (*facing page*). *Upper*: A young queen of *Atta texana* with her first daughter workers. At this early stage of colony development, the adult workers are very small. *Lower*: Super-majors (soldiers) and media workers in the fungus garden of an *Atta cephalotes* nest. The super-majors usually do not participate in fungus garden work. (Photos: Alex Wild.)

In addition, recent genetic evidence contradicts the once widely held perceptions of obligate clonality in the fungal symbionts of leafcutter ants. This research documents "long-lasting horizontal transmission of symbionts between leafcutter taxa on mainland Central and South America and those endemic to Cuba." This suggests that the coevolution of leafcutter ants and their fungal symbionts is not reciprocal. The researchers propose that "a single widespread and sexual fungal symbiont species is engaged in multiple interactions with divergent ant lineages."[21]

3

THE ASCENT OF THE LEAFCUTTERS

Most fungus-growing ant species exist at a level still far from the highest evolutionary grade of superorganismic organization. Lower attines live mostly in relatively small colonies of fewer than 100 to 1,000 individuals, and their nests are inconspicuous and harbor relatively small fungus gardens. Species of the lower attines do not cut and use leaves as the main substrate for their symbiotic fungus, but rather gather a large variety of dead vegetable matter, including bits of leaves, plant seeds, and fruits, as well as insect feces and corpses.[22] Their social organization is relatively simple, with at most only minor polymorphism in minor worker size. It stands in sharp contrast to the leaf-cutting ants of the genera *Acromyrmex* (24 species, 35 subspecies) and *Atta* (15 species). At the extreme, the mature societies of certain *Atta* species are made up of millions of workers inhabiting huge subterranean nest structures with hundreds of interconnected fungus garden chambers. The *Atta* societies represent a benchmark for the spectacular lifestyles "invented" by the ants in the course of their more than 120 million years of evolutionary history. In the following pages, we will focus mainly on this particular genus. The principal life history traits of the various species are very similar across *Atta,* permitting us to present a general picture of their natural history.

The leafcutter ants are of immense importance in tropical and subtropical ecosystems, and they are also major herbivorous pests in cultivated fields through much of Central and South America.[23] For example, in a recent long-term study in the Panamanian rainforest, Rainer Wirth and his colleagues determined that mature *Atta colombica* colonies harvest between 85 and 470 kilograms (dry weight) total plant biomass per colony per year; this corresponds to a harvested leaf area of 835 to 4,550 square meters per year.[24] Such harvesting and processing of enormous amounts of plant material, which is needed for culturing the symbiotic fungus, is possible only by means of cooperation and division of labor among thousands of individuals.

4

LIFE CYCLE OF THE
LEAFCUTTER ANTS

Each of the gigantic *Atta* colonies usually consists of only one queen, the exclusive reproductive individual, and hundreds of thousands or even millions of sterile workers of different sizes and shapes (Plate 15). Each year, mature colonies produce young reproductive females and males, the alates, who depart from their mother colonies on nuptial (mating) flights (Plate 16). The flights of all *Acromyrmex* and *Atta* colonies belonging to the same species and living in the same habitat appear to be synchronized. In *Atta sexdens,* of South America, for example, they take place in the afternoon at any time from the end of October to the middle of December, whereas in *Atta texana,* of the southern United States, they occur at night. Mating itself in *Atta* species takes place high in the air, and since many colonies conduct their nuptial flights during the same time of the day, the probability of outbreeding is high. *Acromyrmex* species also mate on the wing, but copulating pairs often fall to the ground and are immediately surrounded by additional males who struggle for mating access to the

PLATE 15. A queen of an established colony of *Atta cephalotes* from Panama covered by worker ants, who groom and protect her. Note the tiny minims to the near side of the queen and the super-major on the left side of the picture. (Photo: Bert Hölldobler.)

females (see Plate 16). Although mating in *Atta* has never been observed in nature, it has been estimated from sperm counts in the spermatheca of newly mated *Atta sexdens* queens that each queen is inseminated by three to eight males.[25] Such polyandry was later confirmed in studies employing DNA analyses. For example, in *Atta colombica,* the average number of fathers per colony is a bit below three. Owing to variation in shared paternity, the effective paternity frequency in this species is only two.[26] The range of fathers per *Atta*

PLATE 16. *Upper*: A mating swarm of males of *Acromyrmex versicolor* above the Arizonan desert. Females are attracted most likely by a pheromone the males emit and fly into the male swarm to mate. The mating swarms of these leafcutter ants occur after the summer monsoon rains. *Lower left*: A male of *Acromyrmex versicolor*. *Lower right*: A queen of *Acromyrmex versicolor* has fallen to the ground and is surrounded by several males who struggle to mate with her. *Acromyrmex* queens typically mate with several males. (Photos: Alex Wild.)

sexdens colony is between one and five.[27] The mating frequency of *Acromyrmex* queens ranges between one and ten males.[28] In contrast, queens of *Sericomyrmex* and *Trachymyrmex* and those of the lower attine genera appear to be all singly mated.[29]

The biological significance of multiple paternity in leafcutter ants is not entirely clear. Obviously, multiple matings of the queen decreases the average relatedness among the workers in the colony. It has been argued that an increase in genetic diversity may confer an advantage of colony fitness—for instance, in regard to disease resistance.[30] This might be of particular importance in those fungus-growing ant colonies that exhibit vast expansions and exist for many years. Such a body of organic matter underground and such large numbers of ants are very susceptible to parasites and pathogens. An increase in genetically determined defense and resistance mechanisms in the ants obviously is beneficial for colony survival. An enhancement of genetic diversity in fungus-growing ants is especially important in attine ants that cultivate a clonal fungus lineage on which they have been dependent perhaps for millions of years. A long duration of this kind can be expected to lower genetic diversity in the gardens and thus render the cultivar more susceptible to disease, which in turn would require enhanced sanitary defenses in the ants.[31]

Several studies with honeybees provide convincing evidence in support of the hypothesis that multiple mating of the queen improves the colony's vitality and resistance to disease in insect societies. In one study, brood nest temperatures in genetically diverse colonies were found to be more stable than in genetically uniform colonies.[32] Even more important, Thomas Seeley and David Tarpy were able to demonstrate that polyandry improves the colony's resistance to disease.[33] Experimental honeybee colonies were inoculated with spores of the bacterium *Paenibacillus larvae,* which causes the highly virulent disease called American foulbrood. Colonies headed by a multiply inseminated queen had markedly lower disease intensity and higher colony strength relative to colonies headed by a singly inseminated queen. Furthermore, Heather Mattila and Thomas Seeley discovered that "accumulated differences in foraging rates, food storage, and population growth led to impressive boosts" in drone production, winter survival, and colony founding of genetically diverse colonies.[34]

An alternative hypothesis to explain multiple matings is that a queen requires a large lifetime supply of sperm. Leafcutter colonies are as a rule extremely populous and have a life span of ten to fifteen years or even longer. During her lifetime, the queen produces 150 million to 200 million female offspring (female alates and workers). She stores approximately 200 million to 320 million sperm cells in her spermatheca.[35] It can be argued that to obtain an optimal store of sperm that will last for more than a decade, a queen has to mate with multiple males. It has been documented that multiple mating, as expected, does increase the amount of sperm in the spermatheca in *Atta* queens.[36]

Finally, a third competing hypothesis states that high within-colony genetic diversity has a positive effect on worker task efficiency and thus enhances genetic disposition for the development of morphological subcastes in *Atta* colonies.[37] Such partial hardwiring of labor division would be favored by colony-level selection.

Each of these three hypotheses is supported by circumstantial and nonexclusive evidence, and within-colony genetic diversity may in fact have multiple adaptive significance.

After the mating flight, all males die. The sole function of male ants is to provide sperm, which are stored and kept alive for many years in the spermatheca of the queens. Thus, the life span of male ants (which develop from unfertilized eggs and are therefore haploid)

is very short. However, because of the long preservation time of sperm in the queen's internal "sperm bank," males can become fathers many years after they have died. Mortality is also very high for the young queens, especially during the mating flight and immediately after, when the queens have shed their wings and attempt to start new colonies (Plate 17). Out of 13,300 newly founded colonies of *Atta capiguara* in Brazil followed during one study, only 12 were alive three months later. From a start of 3,558 incipient *Atta sexdens* colonies, only 90 (2.5 percent) were alive after three months. In another study, only 10 percent of *Atta cephalotes* colonies survived the first few months after colony foundation.[38]

Before departing on her mating flight, each *Atta* queen packs a small wad of mycelia of the symbiotic fungus into her infrabuccal pocket (cibarium), a cavity located beneath the opening of the esophagus. Following the nuptial flight, the queen casts off her wings and excavates a nest chamber in the soil. This incipient nest consists of a narrow entrance gallery that descends 20 to 30 centimeters to a single chamber about 6 centimeters in length (Figure 1). The queen now spits out the mycelial wad and feeds it with her first eggs, which then serves as an inoculum to start a new fungus garden (Plate 18). By the third day, fresh mycelia have begun to grow, and the queen has laid three to six eggs.[39] At the end of the first month, the brood, now consisting of eggs, larvae, and perhaps pupae, is embedded in the center of a mat of proliferating fungus. During this initial phase of colony foundation, the queen cultivates the fungus garden herself, mainly by fertilizing the garden with fecal liquid. The queen consumes 90 percent of the eggs she lays. When the first larvae hatch, they are also fed with eggs.

PLATE 17 (*facing page*). *Upper*: After mating, the *Acromyrmex* queen sheds her wings. *Lower*: Finding an appropriate place to dig the founding nest chamber is a dangerous activity for the young queen. Here two harvester ant workers (*Aphaenogaster cockerelli*) attack a young *Acromyrmex versicolor* queen. (Photos: Alex Wild.)

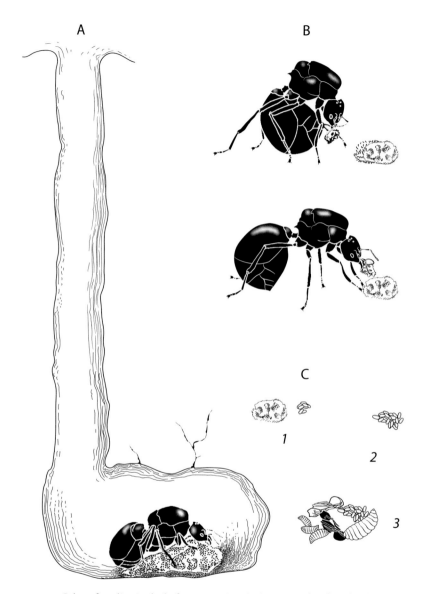

A

B

C

1

2

3

FIGURE 1. Colony founding in the leafcutter ant *Atta*. A: A queen in her first chamber with the beginning fungus garden. B: The queen manures the garden by pulling a hyphal clump free and applying an anal drop to it. C: Three stages in the growth of the garden and of the first brood, which occur simultaneously. (Redrawn by Margaret Nelson, based on an original drawing by Turid Hölldobler-Forsyth.)

PLATE 18. *Upper*: A young queen of *Acromyrmex versicolor* establishing a fungus culture. The fungal inoculum, which she carried in the pouch (infrabuccal pocket) in the mouth cavity, has been expelled and fertilized with freshly laid eggs. *Lower*: In a more advanced stage of colony foundation, the *Acromyrmex versicolor* queen tends larvae in the incipient fungus garden. (Photo: Alex Wild.)

Apparently, the queen does not feed on the initial fungus culture, which is very fragile. If the queen fails to build up a healthy fungus garden, the whole colony-founding process is doomed. Instead, the queen subsists entirely on her own fat-body reserves and by catabolizing her now useless wing muscles.

When the first workers eclose, they begin to feed on the fungus, and they take over the fungus culture activities. The egg-laying rate of the queen now increases. Not all her eggs are viable; some are large trophic eggs formed in the oviduct by the fusion of two or more distinct but malformed eggs. These are given by workers to developing larvae. After a week or so, the young workers open the clogged nest entrance and start foraging in the immediate vicinity. They collect bits of leaves, which they add to the substrate of the fungus culture. By this time, the queen has ceased attending to the brood and fungus garden and has become an "egg-laying machine," a role she will keep for the rest of her long life. The queen is constantly surrounded by workers who groom and feed her with worker-laid trophic eggs. Indeed, the queen is the most precious part of the colony (see Plate 15). She is the reproductive unit of the superorganism. If she dies, the colony is doomed. The workers have assumed all "somatic" duties of the colony: foraging, caring for the fungus garden, raising the brood, extending the nest structures, and defending the colony against predators and competitors.[40]

As the fresh leaves and plant cuttings are brought into the nest, they are cut into smaller and smaller pieces and treated with the ants' fecal liquid before being inserted into the garden substratum (Plates 19 to 24). The ants subsequently pluck tufts of mycelia from other parts of the garden and plant them on newly formed portions of the substratum. The inoculum proliferates swiftly thereafter: the transplanted mycelia grow as much as 13 micrometers per hour. With increasing size, the fungus garden also

PLATE 19. *Upper*: A foundress queen of *Atta vollenweideri*. After nurturing the incipient fungus garden to a certain size, the queen begins to raise her first worker daughters. (Photo: Helga Heilmann.) *Lower*: A foundress queen of *Atta vollenweideri* from Argentina surrounded by her first dwarf worker daughters. (Photo: Bert Hölldobler.)

serves for the partitioning of the large nest cavities excavated from the soil. The larvae and pupae are housed in fungus garden chambers where nurse ants care for the immatures (see Plate 23).

The fungus cultivated by *Atta* and *Acromyrmex* species produces hyphal tip swellings, called gongylidia, which form into densely packed clusters called staphylae (see Plate 22). These

PLATE 20. A worker of *Atta cephalotes* carrying a leaf fragment to the nest. (Photo: Alex Wild.)

PLATE 21. Brood chambers and fungus garden are intimately intertwined, as shown in this view of the fungus garden of a mature *Atta sexdens* colony. Note the diversity of worker subcastes engaged in various activities. (Photo: Bert Hölldobler.)

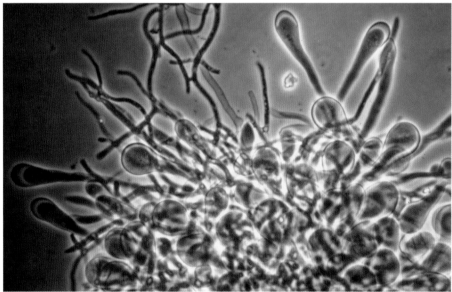

PLATE 22. *Upper*: The minor and media workers process the vegetable material in the fungus garden and tend the fungus. *Lower*: Microscopic photograph of the hyphal swellings, called gongylidia, of the *Atta* fungus garden. (Photos: Bert Hölldobler.)

PLATE 23. *Upper*: The pupal brood chambers in the midst of the fungus garden of an *Atta sexdens* colony. One can see the legs and antennae tightly folded to the still unpigmented body of the immature ant. *Lower*: Inside the brood chamber in the midst of the fungus garden, a pupa of a pigmented super-major (soldier) of *Atta cephalotes* is surrounded and groomed by nurse ants. One can easily recognize an eye, mandibles, and legs of the developing soldier ant. (Photo: Bert Hölldobler.)

aggregates are easily plucked by the ants and eaten or fed to the larvae. The structures are rich in lipids and carbohydrates, while the hyphae are richer in proteins.[41] When given a choice during feeding experiments, *Atta* workers prefer staphylae over hyphae. In addition, they live longer when feeding on staphylae rather than on hyphae.[42] Thus, the staphylae appear to possess the best-balanced blend of nutritional components.

An important nutritional interdependency of the symbiotic fungus and leaf-cutting ants was revealed in a study of the metabolism of plant polysaccharides by the fungus of *Atta sexdens.* It is generally assumed that once the fungus degrades and assimilates cellulose, xylan, pectin, and starch, it is able to mediate the transference of carbon from plant material to the ants. This metabolic integration enables ants to exploit solid plant material not otherwise available to them. The integration primarily involves xylan and starch, both of which support rapid fungal growth. Cellulose, contrary to previous assumptions, seems to be less important, because it is poorly degraded and assimilated by the fungus. Thus, if these biochemical analytical results drawn from laboratory cultures correctly reflect the fungal role in the symbiosis in nature, xylan and starch, not cellulose, are the main leaf polysaccharides contributing to ant nutrition.[43]

The main findings have been substantiated in recent studies in *Atta* and *Acromyrmex,* where it turns out that cellulose is not used as a main energy and carbon source for the fungus-ant association. In fact, strong circumstantial evidence suggests that the

PLATE 24 (*facing page*). *Upper*: Super-major and minor (minim) of the leafcutter ant *Atta cephalotes*. This picture strikingly illustrates the extremes of worker sizes in a single colony. (Photo: Alex Wild.) *Lower*: Members of the smallest worker subcaste (minims) often ride on the transported vegetation during the harvesting operation. These hitchhikers protect the carrier ant from attacks by parasitic phorid flies. (Photo: Bert Hölldobler.)

fungus cannot degrade cellulose at all.[44] Recent studies succeeded in identifying a fungal gene for the degradation of xylan in the fungus gardens of leafcutter ants.[45]

In another study, worker extract has been reported to display high enzymatic activity on starch, maltose, sucrose, and a glycoside. Similar but higher enzymatic activity occurs in larval extract. In particular, the enzymes degrade sucrose, maltose, and laminarin, the latter a hemicellulose cell wall component of plants. Some variation in enzymatic activities occurs in the extract from symbiotic fungi of different *Acromyrmex* species. In the study, the fungal extract of *Acromyrmex subterraneus* was most active on laminarin, xylan, and cellulose, while the fungal extract of *Acromyrmex crassispinus* was most active on laminarin, starch, maltose, and sucrose.[46]

These results, especially those regarding the degradation of laminarin and cellulose, seem to contradict earlier findings. That ant extracts can degrade the plant macromolecule laminarin is especially problematic. The difficulty may be resolved by the fact that fungal enzymes pass through the ant gut. Most likely, the enzymes detected in ant larval extract derived in part from the consumed fungus.

In any case, the fungus is not the only source of nutrition for leaf-cutting workers. In the laboratory at least, *Atta* and *Acromyrmex* workers fed directly on plant sap. The sap appears to be the "fuel" that provides the energy for the leafcutters and harvest transporters. In fact, intake of sap appears to be crucial to the workers, because in laboratory experiments, only 5 percent of the energy requirements were met by ingestion of the contents of fungal staphylae.[47] In contrast, the larvae are able to subsist and grow entirely on the staphylae. The queen appears to obtain a substantial part of her food from trophic eggs laid by workers and fed to her at frequent intervals.

5

THE *ATTA* CASTE SYSTEM

The growth of an incipient colony is slow in the first two years. During the next three years, it accelerates quickly and tapers off as the colony starts to produce winged males and queens. The ultimate colony size reached by *Atta* is enormous: the number of workers in a single colony has been estimated at 1 to 2.5 million in *Atta colombica,* 3.5 million in *Atta laevigata,* 5 to 8 million in *Atta sexdens,* and 4 to 7 million in *Atta vollenweideri.*[48]

Among the fungus-growing ants, only species of the two leaf-cutting genera, *Acromyrmex* and *Atta,* have highly polymorphic workers, with strong differences in size and anatomical proportion. This remarkable polymorphism is reflected in the complex division of labor exhibited within the colonies. A rich literature exists on various aspects of division of labor in *Atta* species. Most of the studies are in agreement concerning the major patterns that characterize division of labor in *Atta* colonies.[49] The following account is based on the labor system of *Atta cephalotes* and *Atta sexdens.*[50]

Atta leaf-cutting ants have a broad array of physical subcastes in the worker groups (Figure 2). In *Atta sexdens,* for example, the head width varies 8-fold and the dry weight 200-fold from the smallest minor

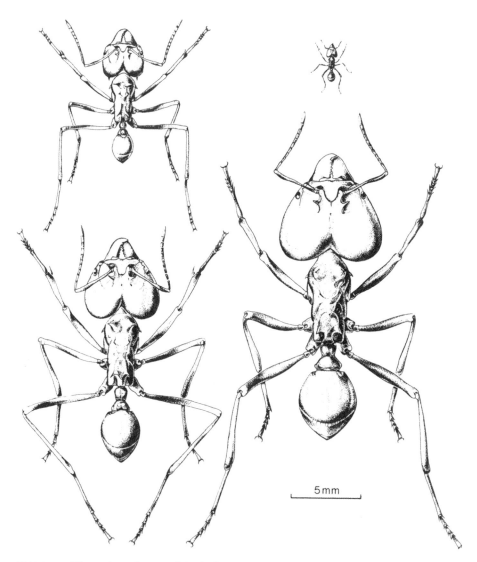

5mm

FIGURE 2. The worker subcastes of the leafcutter ant *Atta laevigata.* Species of the genus *Atta* have the most complex systems of division of labor known in ants. (Drawing by Turid Hölldobler-Forsyth from G. Oster and E. O. Wilson, *Caste and Ecology in the Social Insects* [Princeton, NJ: Princeton University Press, 1978].)

workers to the huge major workers (see Plate 24). However, developing colonies, started by a single queen, have a nearly uniform size frequency distributed across a relatively narrow head-width range of 0.8 to 1.6 millimeters (see Plates 14 and 19). The reason for this restriction is an experimentally demonstrated necessity: workers in the span of 0.8 to 1.0 millimeter are required as gardeners of the symbiotic fungus, whereas workers with a head width of 1.6 millimeters are the smallest that can cut vegetation of average toughness. The combined range (0.8 to 1.6 millimeters) also embraces the worker size groups most involved in brood care. Thus, the queen produces about the maximum number of individuals who together can perform all the essential colony tasks. As the colony continues growing, the worker size variation broadens in both directions, to head width 0.7 millimeter or slightly less at the lower end and to more than 5 millimeters at the upper end, while the frequency distribution becomes more sharply peaked and strongly skewed to the larger-size classes. This complex caste system reflects the division of labor in *Atta,* which is closely adapted to the collection and processing of fresh vegetation for fungal substrate and to the culturing of the fungus.

The *Atta* workers organize the gardening operation in the form of an assembly line. The most frequent size group among foragers, at the start of the line, consists of workers with a head width of 2.0 to 2.2 millimeters. At the end of the line, the care of the delicate fungal hyphae requires very small workers, a task filled within the nest by workers with a head width of predominantly 0.8 millimeter. The intervening steps in gardening are conducted by workers of graded intermediate size (see Plates 20 and 22).

After the returning foragers (Figure 3, activity 1) drop the pieces of vegetation onto the floor of a nest chamber, the pieces are picked

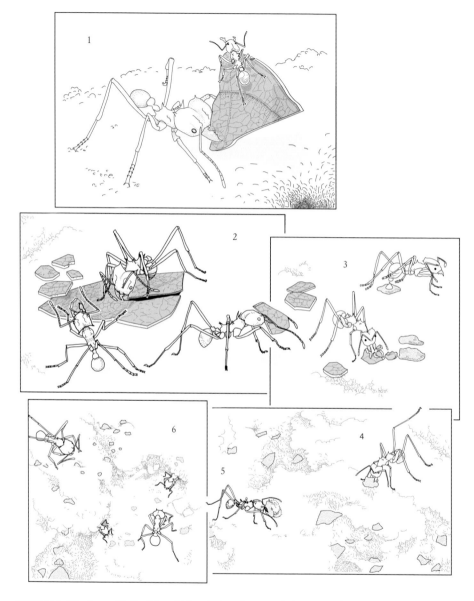

FIGURE 3. The "assembly line" by which colonies of *Atta cephalotes* create fungus garden with fresh-cut leaves and other vegetation. (Illustration by Margaret Nelson.)

up by workers of slightly smaller size, who clip them into fragments about 1 to 2 millimeters across (activity 2). Within minutes, still smaller ants take over, crush and mold the fragments into moist pellets, add fecal droplets (activity 3), and carefully insert them into a mass of similar material (activity 4). Next, workers even smaller than those just described pluck loose strands of fungus from places of dense growth and plant them on the newly constructed surfaces (activity 5). Finally, the very smallest and most abundant workers patrol the beds of fungal strands, delicately probing them with their antennae, licking their surfaces, and plucking out spores and hyphae of alien species of mold (activity 6).

Superimposed on this division of labor, which is based on anatomical worker subcastes, is age polyethism: young workers of most subcastes perform tasks inside the nest, and older workers tend to be involved in tasks outside the nest. This distinction is strikingly illustrated by the smallest worker subcastes (the so-called minim workers), which inside the nest tend the fungus and small brood, but which can also be seen at the harvesting site, even though they are unable to cut and carry leaf fragments. Many of them do not walk back to the nest on their own, but ride ("hitchhike") on the leaf fragments being carried to the nest (see Plates 24 and 37). Most likely, these individuals are older minim workers that defend the leaf carriers from attacks by parasitic phorid flies, which attempt to lay eggs on the ants' bodies. But the role of the hitchhikers might be more diverse. Timothy Linksvayer and his colleagues observed hitchhikers in *Atta cephalotes* colonies in La Selva, Costa Rica, which foraged at night when parasitic phorid flies were not flying. From their detailed observation of the behavior of the hitchhikers they suggest that hitchhikers also clean the transported leaf fragments

PLATE 25. Army ants of the genus *Nomamyrmex* frequently attack leafcutter ant colonies. Here two *Atta* workers in the rainforest of Panama attack a scout of *Nomamyrmex esenbeckii*. (Photo: Alex Wild.)

PLATE 26. *Upper*: The head of a super-major (soldier) *Atta cephalotes*. The sharp mandibles powered by massive muscles can easily cut human skin. (Photo: Alex Wild.) *Lower*: A super-major (soldier) of *Atta cephalotes*. (Photo: Bert Hölldobler.)

from fungal contaminations and other harmful microorganisms. A subsequent experimental study with *Atta sexdens* and *Atta laevigata* led by E. H. M. Vieira-Neto strongly supports this suggestion.[51]

Most size groups of *Atta* colonies engage in defense of the colony, but again, it is the older workers that most likely attack intruders and defend territories (Plate 25). At the same time, colony defense is organized to some extent according to worker size. There is, for example, a true soldier caste. These extremely large majors have sharp mandibles powered by massive adductor muscles (Plate 26). They are especially adept at repelling large enemies, especially vertebrates. The differential involvement of worker castes in colony defense has been well documented in a study of *Atta laevigata*. When the colony is threatened by a potential vertebrate predator, mostly the gigantic soldiers are recruited. However, when a colony has to defend its nest or foraging area against conspecific or interspecific ant competitors, mainly smaller worker castes respond. These are more numerous and more suitable in territorial combat with enemy ants.[52] Similar results have been reported from studies of the grass-cutting ant *Atta capiguara*. In this species, minor workers responded most readily to alarm pheromones experimentally released from the mandibular glands of *Atta capiguara* workers near the foraging trail. The response was strongest when the pheromone was released close to the trail.[53] Foragers transporting grass fragments did not respond at all. However, in *Atta colombica*, whose nests are raided by the ecitonine army ant *Nomamyrmex esenbeckii*, workers recruited mainly the majors (soldiers) as a specific defensive response to army ant attacks.[54]

6

HARVESTING VEGETATION

Variation in body size may also be important in the harvesting behavior of *Atta* workers.[55] A leaf-cutting forager typically harvests leaf pieces the mass of which corresponds with her body size. This could be a result of the leaf-cutting behavior. During cutting, a worker usually anchors her hind legs on the leaf edge and slowly pivots around the body axis, pushing the cutting mandible through the leaf tissue (Plates 27 to 29). In this way, the fragment size is correlated with the body size of the cutter. In other studies, however, no relationship between the length of the legs of *Atta cephalotes* ants and the cut curvature of the leaf fragment has been found.[56] Instead, the angle between head and thorax can be changed by the cutting ant, allowing it considerable flexibility. Hence, the fragment size is not a simple function of the legs acting as a pivot. It can also be argued that the leaf-cutting foragers do not directly assess fragment mass while cutting, but use leaf toughness as an indirect measure for adjusting the size of the cut fragments. Thus, although ants cannot cut larger pieces than their overall body size permits (unless they move along the cutting edge), they are able to change their posture in order to cut smaller leaf fragments. If leafcutters encounter soft fruits, they will eagerly engage in

PLATE 27. In the first step of vegetation processing, an *Atta sexdens* forager shears a leaf fragment from the tree at the harvesting site. Only one mandible functions as "cutting knife," leaving the other to serve as a "pacemaker." The leafcutter ant also uses the foot of the foreleg on the side of the cutting mandible to pull the cut edge of the leaf fragment upward. This motion evidently increases the stiffness of the leaf and aids in the cutting process. Note how the ant explores the leaf surface to which the cut is directed with her right antenna. (Photo: Bert Hölldobler.)

cutting chunks from the fruit and carry these to their nest. Because of their larger mandibles, the large workers usually cut big pieces out of the fruit. This has been reported from *Atta laevigata*, where size differences among the workers lead to task partitioning whereby the large workers are the fruit cutters and medium-sized workers serve as fruit carriers. It obviously increases the total harvest transport rate if larger loads per trip are transported.[57]

During cutting, the two mandibles of *Atta* workers play different

PLATE 28. *Upper*: A leafcutter ant of *Atta cephalotes* about to finish the circular cut out of a leaf fragment. (Photo: Alex Wild.) *Lower*: Harvesting activity of *Atta* leafcutters in the tree canopy seen against the blue sky light. (Photo: Christian Ziegler.)

roles. While one mandible actively moves, the other remains almost fixed and serves as the cutting jaw (see Plates 27 and 29). The steps in one full bite are as follows (Figures 4 and 5). The motile mandible is opened and anchored with its tip to the leaf tissue. The cutting mandible is not opened, but held steady. During the opening of the motile mandible, the cutting jaw is pushed against the leaf by lateral head movements. Next, the motile mandible is closed, pulling the cutting jaw further against the leaf and lengthening the incision. In this phase, the motile mandible also moves deeper into the leaf surface, thus preparing the way for the cutting jaw. As soon as both jaws meet, the cycle starts again. Thus, one jaw functions as "cutting knife" and the other one as "pacemaker." But there is no "sidedness": either right or left jaw can function as cutting knife, depending on the direction in which the leaf fragment is cut.

Leafcutters often stridulate while they are cutting. A number of workers cutting leaf fragments raise and lower their gasters in a motion identical to that performed by *Atta* workers when producing sound (see Figures 4 and 5). The sound comes from a stridulatory organ, composed of a cuticular file on the first gastric tergite and a scraper situated on the postpetiole. By rubbing the file against the scraper, the ants produce audible vibrations.[58] The analysis of the temporal relation between mandible movements and stridulation, made by videotaping the cutting behavior and simultaneously recording the vibrational signals from the leaf surface with laser vibrometry, has revealed that stridulation occurs most often when the cutting mandible is moved through the plant tissue (see Figure 5). The stridulation generates complex vibrations of the mandibles, which give the mandibles some of the properties

PLATE 29 (*facing page*). *Upper: Acromyrmex lundi* leafcutter from Argentina slices a piece from a leaf. *Lower: Acromyrmex coronatus* from Panama cut circular fragments from a leaf. (Photos: Alex Wild.)

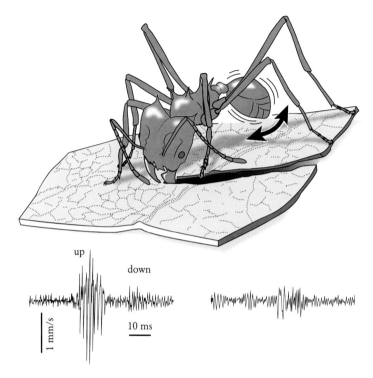

up

down

1 mm/s

10 ms

FIGURE 4. The cutting of a leaf by a leafcutter ant. The worker vibrates her abdomen, producing a stridulation sound, as she scissors her way along. The substrate-borne vibrations facilitate the cutting process but, more important, serve as a close-range recruitment signal. They attract nestmates from the vicinity to help in harvesting a high-quality leaf. The stridulation signals were recorded by laser-Doppler vibrometry (as velocity of the leaf's vibrations). *Lower left*: Substrate-borne vibrations transmitted mostly through the mandibles while cutting. *Lower right*: Vibrations transmitted onto the substrate through the legs when the mandibles do not touch the leaf. (Illustrations by Margaret Nelson, based on F. Roces, J. Tautz, and B. Hölldobler, "Stridulation in leaf-cutting ants: short-range recruitment through plant-borne vibrations," *Naturwissenschaften* 80[11]: 521–524 [1993].)

of a vibratome (the vibrating knife of a microtome). Indeed, when the cutting process was experimentally simulated, it turned out that the vibrating mandible reduces the force fluctuations that inevitably occur when material is being cut. Thus, stridulatory vibrations facilitate a smoother cut through tender leaf tissue.[59]

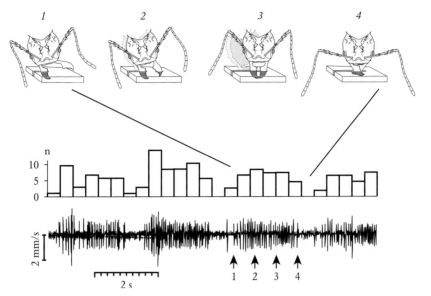

FIGURE 5. *Upper*: Mandible and head movements during one cut into a tender leaf. *Lower*: Stridulation during four bites. The histogram shows the number of chirps counted through a span of 400 milliseconds. The trace underneath depicts an original laser vibrometry of stridulation on a leaf. The arrows denote the temporal occurrence of the four cutting stages shown above. (Redrawn by Margaret Nelson, based on J. Tautz, F. Roces, and B. Hölldobler, "Stridulation in leaf-cutting ants: short-range recruitment through plant-borne vibrations," *Naturwissenschaften* 80[11]: 521–524 [1993].)

Cutting fragments out of leaves requires powerful mandible muscles. Accordingly, the mandibular muscles in *Atta* make up more than 50 percent of their head capsule mass, or more than 25 percent of the entire body mass.[60] Leaf cutting is also an extraordinarily intense behavior energetically. The leaf-cutting metabolic rate, which has been determined in an extremely sensitive flow-through respirometry system, is dramatically above both the standard rate and the post-cutting locomotor metabolic rate. The aerobic scope of leaf cutting has been determined to be within the same range as that of flying insects, which are among the most metabolically active of all animals.

The mandibular energetics of leaf cutting therefore probably plays an important role in the ants' load size selection and foraging efficiency at both the individual and the colony levels (Plates 30 to 33).

During the past two decades, numerous papers have been published addressing the question of load size selection in leaf-cutting ants. It is beyond the scope of this presentation to review the many different and sometimes contradictory results obtained. Obviously, the parameters that affect load size are numerous. Although, as noted, there is a correspondence between the size of the leaf-cutting worker and the leaf fragment size (area) to be cut, fragment size is not always the best parameter to determine load size (mass). The reason is that the mass of a fragment depends on leaf mass per unit surface area as well as on leaf fragment volume. Foragers of *Atta cephalotes* and *Atta texana* do not tend to adjust leaf-cutting behavior as a function of leaf density. Instead, workers of different sizes tend to cut leaves of different densities.[61] Similar patterns of forager polymorphism and resource matching have been found in other *Atta* species. On the other hand, several additional independent studies have revealed that the denser the leaf, the smaller the fragments.[62]

The mass of a leaf fragment being transported also affects the running speed of the carrier ant, and both parameters (load mass and retrieval time) affect the colony's rate of vegetable material intake.[63] The slower speeds of the workers carrying heavier fragments may not negatively affect intake rates, because the yield per delivery is increased. But extended travel time, owing to heavier loads, may have other detrimental consequences. For example, the transfer of

PLATE 30 (*facing page*). *Upper*: Worker of *Atta sexdens* loading a leaf fragment into the "umbrella" position to carry it to the nest. (Photo: Bert Hölldobler.) *Lower*: Leaf pickups by *Atta* leafcutter ants. Leaf fragments dropped to the ground from the harvesting site up in the tree are picked up by the carriers, who transport the harvest to the nest. (Photo: Christian Ziegler.)

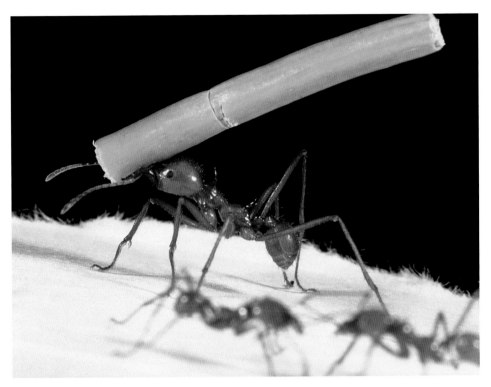

PLATE 31. Vegetable fragments harvested by *Atta* leafcutters are transported to the nest. (Photos: Bert Hölldobler.)

PLATE 32. The foraging columns of *Atta* species are typically crowded with leaf carriers. (Photo: Hubert Herz.)

information about the food resource to the colony can be delayed and therefore the speed and intensity of recruitment weakened.[64]

Short travel time therefore appears to be an asset in the foraging system of leaf-cutting ants, thus favoring load sizes that have a minimal effect on the running speed of leaf carriers. In any case, individual maximization models, often usefully applied to solitary foraging animals, fail to explain fragment selection by *Atta* foragers.[65] In fact,

PLATE 33. Even when not used, the long trunk routes of *Atta* leafcutter ants are distinctly visible, being kept clear of vegetation by road workers. (Photo: Herbert Herz.)

it may well be that small loads are rate maximizing, but at the level of the colony rather than of the individual worker. Indeed, fragment size might be influenced by many factors, including size of the worker ant, energetic cost of cutting, density (mass) of the leaf, need to rapidly transfer foraging information to the colony, distance and quality of the harvesting site, and "handling cost," which most likely increases with fragment size.

Finally, here is one other aspect concerning harvest selection in leafcutter ants. A study by Rainer Wirth and his co-workers demonstrated that foragers of *Atta colombica* prefer to harvest leaves from drought-stressed plants, and within individual plants the ants harvested significantly more leaf fragments from stressed leaves than from vigorously thriving leaves. Further analysis suggests that the ants are attracted to stressed leaves because these contain higher concentrations of the amino acid proline and carbohydrates, such as sugars.[66]

In almost every respect, size matters in the division of labor in *Atta* societies. Especially marked is the immense size and anatomical difference between workers and the gigantic queen, which reaches its extreme in the enormous ovaries of the queen when compared with the "degenerate" ovaries of workers. The queen is the sole reproductive individual in the society.

Additional labor specialization is achieved by programmed shifts with aging. At least three of the four physical castes of *Atta sexdens*, for example, pass through changes of behavior with aging. Although caste and division of labor in this and other *Atta* species are very complex in comparison with other ant systems, they are derived from surprisingly elementary processes of increased size variation, allometry, and alloethism. In fact, ant species in general and *Atta* species in particular have been remarkably restrained in the elaboration of their castes. They have relied on a single rule of deformation to create physical castes, which translates into a single allometric curve for any pair of specific dimensions, such as head width versus pronotal width. Hence, the *Atta* species have not evolved anywhere close to the conceivable limit. There are far more tasks than castes: by the first crude estimate, seven castes cover a total of twenty to thirty tasks. Also, one can discern another important phenomenon in *Atta* species that constrains the elaboration of physical castes: polyethism has evolved further than polymorphism. In the course of evolution, *Atta* created its division of labor primarily by greatly expanding the size variation of the workers while adding a moderate amount of allometry and a relatively much greater amount of alloethism.[67]

Alloethism is the regular change in a particular category of behavior as a function of worker size. It stands in close relationship to the phenomenon of task partitioning: "A task can be said to be partitioned when it is split into two or more sequential stages so that material is passed from one worker to another."[68] This phenomenon is well known to myrmecologists from a number of different species in various contexts. It includes "bucket-brigade" harvesting in leaf-cutting ants, in which leaves are cut by some workers and dropped to the ground for further fragmentation (see Plate 30). The material is then transported to the nest by other workers for varying distances along trunk trails, until the harvest material reaches the nest[69] (see Plates 32 and 33). In some *Atta* species, such as *Atta colombica,* the leaf carriers establish one or more caches along the trail. In others, including *Atta vollenweideri,* the leaves are dropped haphazardly along the trail at variable distances. In grass-cutting ants (*Atta vollenweideri*), grass fragments are harvested and transported to the nest for distances of up to 150 meters along well-established trunk trails.[70] Cutting and transporting of fragments are distinct activities, often performed by separate workers differing in body size. Because leaf cutting is a much more energetically intense activity than transport,[71] colonies can be expected to allocate larger workers to this task. This body size effect is less obvious when the harvest site is located very close to the nest; moreover, no physical trail is present. In such situations, the cutter often carries the grass leaf fragment to the nest alone. However, on long foraging trails, the workers form transport chains composed of two to five carriers per grass fragment. As a rule, the first carriers cover only a short distance before dropping the fragments. Sometimes cutters participate in this first phase of harvest transportation but after dropping their load usually return to the harvesting patch. The last carriers cover the longest distance. Furthermore, the probability of dropping the carried leaf fragment is independent of both worker size and load size.

What are the advantages of this kind of chain transport in *Atta vollenweideri*? Maximization of load transportation has been proposed for those leaf-cutting ant species that employ caches on the ground for transfer of harvested leaf fragments to the nest. However, the empirical data do not always support the assumptions.[72]

Carl Anderson and his colleagues have discussed several advantages and disadvantages of a chain transport of harvest to nests, as in *Atta vollenweideri,* where the last carriers cover the longest distance.[73] The researchers argue that such task partitioning can be expected to enhance the work efficiency of individuals, because workers are more likely to become specialists when deployed sequentially. As a consequence, the colony's overall rate of resource retrieval should be higher. But again, the empirical data do not entirely support these theoretical considerations.

Finally, Jacqueline Röschard and Flavio Roces have proposed a second hypothesis: that the transport chains of *Atta vollenweideri* accelerate transfer of information about the plant species and food quality of the harvest.[74] They argue that the dropping of fragments on the trail allows cutting workers to quickly return to their tasks. In addition, moving along short trail sections during foraging facilitates reinforcement of the trail pheromone markings, which in turn leads to a faster recruitment of a foraging force and subsequent monopolization of the harvesting site. In addition, fragments dropped on the trail may serve as information signals. For example, outgoing foragers could obtain information about the resources actually being harvested. If this "information transfer hypothesis" is correct, transport chains can be expected to occur more frequently under conditions in which the information is valuable, for instance, on discovery of high-quality resources or when the colony is harvest deprived.[75] Röschard and Roces have obtained evidence supporting this conjecture.[76]

For example, field experiments have revealed that when plant fragments of high quality are presented at selected foraging sites, transport chains occur more frequently, and independent of fragment sizes. In addition, high-quality fragments are transferred from one carrier to another after shorter transport travel. More chains and more segments in the chains are responses to an increase in the attractiveness of the load, allowing the first carriers to return quickly to the foraging site. These results, as predicted, suggest that transport chains increase the information flow at the colony level. Additional data obtained with fragments of the same quality but different size do not support the hypothesis that transport chains enhance the economic load carriage at the individual level, as previously suggested.[77]

Overall, leaf quality is an important influence in recruitment and harvesting intensity in leaf-cutting ants. Its parameters include leaf tenderness, nutrient contents, and the presence and quantity of secondary plant chemicals. In one experiment, harvest preference in *Atta cephalotes* was tested by offering the ants fresh leaves of forty-nine woody plant species from a tropical deciduous forest in Costa Rica. Leaf protein content was positively correlated with the number of fragments cut, while secondary chemistry and nutrient availability interacted in determining the attractiveness of plant material to the ant foragers.[78] In another, related study, young, tender leaves of the tropical legume *Inga edulis* were found to be more loaded with secondary chemicals and contain fewer nutrients than mature leaves, but the latter are three times tougher and thus harder to cut. The investigators concluded that the quality of the colony's habitat likely determines whether a colony will harvest more of the less suitable leaves. Ants that locate and harvest from highly suitable host plants avoid *Inga edulis,* while those in poorer habitats accept the legume but, because of the toughness of this plant's older leaves, mostly harvest the otherwise less suitable young leaves.[79] The diversity of the leaf fragments harvested by one *Atta colombica* colony during one year is beautifully depicted in the frontispiece photograph by Christian Ziegler.

Several lines of evidence thus suggest that harvest preferences in *Atta* leafcutter colonies are determined by trade-offs between several parameters. Furthermore, they depend not only on particular leaf traits but also on the properties of the ecosystem as a whole. Comparative bioassays focusing on a small set of parameters do not, in these ants, capture the complex multivariate picture of harvest selection.[80]

7

COMMUNICATION IN *ATTA*

The highly organized cooperative foraging of the *Atta* leafcutters depends on information transfer and social communication. Much of this information transfer occurs on their harvesting routes. Leaf-cutting ants are famous for their extended and persistent foraging trails (see Plates 32 and 33). These durable routes are very obvious to the human eye. They lead masses of foragers to and from harvesting sites, which are either mostly the canopy of trees, a specialization of *Atta cephalotes, Atta colombica,* and *Atta sexdens,* or patches of savanna grass, the target, for example, of the grass-cutting *Atta vollenweideri.* Early behavioral experiments indicated that the foraging trails are chemically marked with secretions from the ants' poison gland sacs.[81] It has been suggested that this trail pheromone contains at least two functional components—one volatile, which serves as a recruitment signal, and the other much less volatile, which functions as a long-lasting orientation cue. Many chemical and behavioral details of the *Atta* poison gland contents and the response to them remain to be elucidated, but some important aspects with respect to pheromonal communication in foraging behavior have been analyzed.[82]

The volatile recruitment component of some *Atta* species was the first ant trail pheromone whose chemical structure was identified.[83] This compound, methyl 4-methylpyrrole-2-carboxylate (MMPC), functions as a recruitment trail pheromone in all *Atta* species except *Atta sexdens,* whose main recruitment trail pheromone component is 3-ethyl-2,5-dimethylpyrazine (EDMP).[84] *Atta* workers in laboratory colonies readily respond to trails drawn with small amounts of these substances. They follow these trails through all the twists and turns chosen by the experimenter. The potency of MMPC is quite amazing: 1 milligram of this substance is theoretically sufficient to draw a trail that foragers of *Atta texana* and *Atta cephalotes* will follow three times around Earth's circumference.[85] And that record has recently been broken in the case of *Atta vollenweideri:* 1 milligram of this trail pheromone would be enough to lay a trail sixty times around the planet, with approximately 50 percent of the foragers of the grass-cutting ant still following.[86]

In fact, several behavioral studies revealed a rich diversity of scent-guided behavior and astounding odor sensitivity in leafcutter ants. Not surprisingly, the olfactory system in attine ants is extremely well developed. Generally in insects, the olfactory pathway begins with the antennae, where the olfactory sensory receptors are located (Plate 34). These sensory neurons carry the information about the odorant molecules to the antennal lobes, which are part of the brain (deutocerebral brain). The antennal lobe is composed of densely packed glomeruli, where sensory neurons connect with projection neurons. The latter transmit information to the higher brain centers (such as the so-called mushroom bodies). The number of glomeruli in the antennal lobe is a good indication of the fine-tuning capacity of the olfactory system. For example, the fruit fly *Drosophila* has only 43 glomeruli in each antennal lobe, whereas all attine ants investigated have no fewer than 257 glomeruli in each

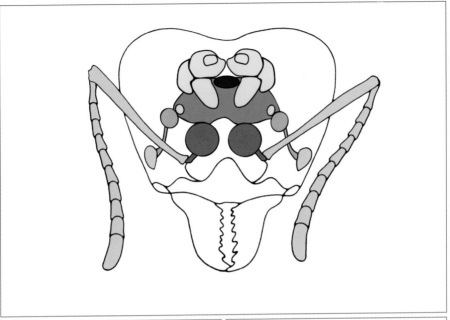

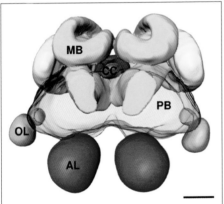

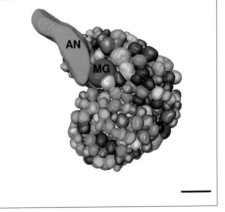

PLATE 34. Schematic illustrations of the brain of an *Atta* worker ant, and its connections to the main sensory organs in the head, the antennae and the eyes. *Upper and lower left*: A three-dimensional reconstruction of the brain of an *Atta* worker, based on microdissections. AL (red): Antennal lobes. OL: Optical lobes (note that the large lamina connecting the eye with the optical lobe is not shown). PB (light blue): Protocerebrum. CC (blue): Central complex. MB (yellow): Mushroom body. *Lower right*: Close-up of antennal lobe with glomeruli. AN (yellow): Antennal nerve. MG (red): Macroglomerulus. (Reconstruction and illustrations: Christina Kelber.)

lobe. As Christina Kelber and her collaborators beautifully demonstrated in a comparative neuroanatomical study, the highest number of 630 is found in the lower attine species *Apterostigma mayri*. In leafcutter ants, the number of glomeruli varies and ranges, depending on species, between 336 to 459. In addition, the larger workers of leafcutter ants possess an extremely large glomerulus (macroglomerulus) near the antennal nerve entrance in each lobe. Circumstantial evidence suggests that this macroglomerulus is involved in processing the trail pheromone information (Plate 34).[87]

Some people attending our public lectures often ask whether ants have a brain, or they wonder "how ants can do all these remarkable feats without a brain." This misunderstanding may derive in part from a sentence in the wonderful book *The Lives of a Cell*, published by Lewis Thomas in 1974, which states, "A solitary ant, afield, cannot be considered to have much of anything on his mind; indeed, with only a few neurons strung together by fibers, he can't be imagined to have a mind at all, much less a thought." Lewis Thomas erred at least on two points. The ant he is talking about is not a "he" but a "she," because all ant societies comprise female members only. Males, as we already discussed, usually have a short life, and their sole function is to deliver sperm into the sperm pocket in the queens' bodies. More important, there are not merely "a few neurons strung together by fibers." Instead, there are close to one million neurons interconnected by axons and dendrites and organized in a nervous system and a remarkable structured brain as shown in Plate 34. As we just noted, particularly complex is the perception and processing of chemical signals in the brain.

The pheromonal markings of the long-distance foraging routes are continuously reinforced by the foragers. However, the fine-tuning of their deposition and the recruitment that results depend on a number of parameters, including the quality of the food and the need of the colony's fungus for new vegetation.[88] The trail phero-mone and other markings also appear to affect the attractiveness of the food source, hence stimulating harvesting activities such as cut-ting and transport of leaf fragments away from the harvesting site.[89] Trail pheromones are used to mark not only the major trunk routes but also the tree branches and twigs frequented by the ants. As a result, foragers continuously perceive the chemical trail signal. Any additional signal that mediates short-range recruitment at the forag-ing site would be most effective if transmitted through a different sensory channel. In fact, just such a superimposed mechanical signal has been found.[90] Most harvesting by leaf-cutting ants as a whole occurs in the canopies of trees. Here one can often observe that col-lectives of ants cut fragments out of particular leaves until nothing is left except a few leaf veins, while other leaves nearby remain almost untouched (Plates 35 and 36). It appears that those leaves intensely frequented by foragers are more desirable than other leaves, perhaps because they are more tender, or richer with sugars and amino acids, or less loaded with secondary plant compounds. Ants are able to summon nestmate foragers to these higher-quality leaves by employ-ing special short-range recruitment signals. The process is as follows. A number of *Atta* workers cutting leaf fragments produce stridu-latory sounds. By employing laser-Doppler vibrometry, researchers have been able to record the signals transmitted by the ants onto the leaf surface (see Figure 4). When ants were offered leaves of differ-ent quality, the proportion of workers that stridulated during cut-ting differed markedly. Significantly more ants stridulated when tender leaves instead of thick leaves were offered. When the quality of the two kinds of leaves was enhanced by a sugar coating, almost all workers that were cutting also stridulated, regardless of differ-ences in the physical properties of the material being cut. These observations suggest that the production of stridulatory vibrations

PLATE 35. Harvesting site of *Atta* leafcutters in Panama. Some leaves are entirely cut down; others stayed intact. (Photo: Bert Hölldobler.)

PLATE 36. Sapling of *Ochroma pyramidale* attacked by *Atta colombica* in Panama. Of certain leaves, only a few hardened leaf veins were left behind by the leafcutter ants, while other leaves remained untouched. (Photo: Hubert Herz.)

is affected by the quality of the leaves, and leaf-cutting foragers use the sound to communicate leaf quality to their nearby nestmates.

In the 1960s, Hubert Markl demonstrated with a series of ingenious experiments that leafcutter ants do not respond to airborne components of the stridulation sound, but they are highly sensitive to vibrations propagated through the substrate.[91] On the basis of these important findings, in subsequent experiments conducted by Flavio Roces and his collaborators, *Atta* workers on their way to the foraging site were given a choice between a vibrating twig and a silent one. When given a choice, more *Atta* foragers respond to recruitment pheromones than to substrate-borne stridulatory vibrations without pheromones. But the effectiveness of the recruitment pheromone is significantly enhanced when it is combined with the vibrational signal. Under natural conditions, nearby workers respond to the stridulatory vibrations transmitted through the plant material by orienting toward the source of the vibrations and subsequently joining in leaf cutting.[92]

The response of the ants to stridulatory signals is context specific. Workers of *Atta sexdens* stridulate as an alarm signal during nest defense. Because stridulation also mechanically facilitates the leaf-cutting process, it is tempting to suppose that leaf cutting was the first function of stridulation and that its employment in communication is a derived trait in evolution. However, subsequent studies have provided circumstantial evidence that the reverse is true: facilitation of cutting by the vibrations is more likely an auxiliary benefit emerging from the communication process.[93]

Leaf-cutting ants also stridulate frequently during nest building, particularly while manipulating soil particles with their mandibles. The stridulatory vibrations might serve in close-range recruitment to summon help from nestmates. But it is also possible that the vibrations simultaneously enhance excavation by acting like a vibrating pneumatic drill.[94]

There is yet one more context in which leafcutter stridulation serves in communication. Minim workers, the smallest worker subcaste, often ride on the leaf fragments being carried by other ants to the nest (see Figure 3, activity 1, and Plates 24 and 37). These tiny guard ants defend the leaf carriers from attack by parasitic phorid flies that try to oviposit on their sisters' bodies. It has been shown that leaf carriers communicate to the hitchhikers their readiness to load up and walk home by means of plant-borne stridulatory vibrations. The stridulatory vibrations produced by the carrier in this initial transport phase seem to attract the tiny minims, who mount the carrier and leaf fragment.[95]

PLATE 37. An *Atta cephalotes* worker carries a leaf fragment with minim nestmates riding as hitchhikers on the transported leaf. The little ants are attracted to mount the carrier by stridulation vibrations that their larger nestmate produces when loading up. Hitchhiker ants protect the defenseless carrier from attacks by parasitic phorid flies. (Photo: Alex Wild.)

Leaf-cutting workers conspicuously stridulate when prevented from moving freely, whether trapped in a partial cave-in or held by an enemy ant. The substrate-borne stridulatory vibrations elicit a close-range alarm effect in nestmates.[96] Ants are attracted by the signal, and they start digging in an attempt to free the nestmate, or they attack the enemy ant that grasps her. Under natural conditions, such rescue signaling is usually multimodal: the mechanical stridulation is an important synergist of alarm pheromones. However, in massive aggressive interactions, with dozens or hundreds of workers involved in a melee, as in territorial defense, pheromones combined with defensive secretions are much more important.

Atta and *Acromyrmex* workers, like those of most other ant species, produce alarm pheromones in their mandibular glands (these glands are attached to the ant's jaws). In fact, the mandibular gland pheromone in *Atta sexdens* was one of the very first such pheromones chemically and behaviorally characterized.[97] Adolf Butenandt and his collaborators identified citral as the main compound in the mandibular gland secretions. These authors noted the relatively large size of the mandibular glands in the largest worker subcastes (soldiers), and they estimated that the glands occupy one-fifth of the volume of the head capsule. In behavioral tests, they demonstrated that the secretions have an alarm and repellent function. Subsequent studies by other researchers came to different results: although they identified citral, geranial, neral, and many other compounds, the effective alarm pheromone was determined to be 4-methyl-3-heptanone.[98] Later studies revealed that smaller workers, who are primarily active inside the nest, have mandibular gland secretions containing largely 4-methyl-3-heptanone, whereas secretions of larger workers engaged primarily outside the nest contain mainly citral.[99] These interesting findings are in accordance with the previously discussed division of labor based on physical worker subcastes and strongly suggest a context-specific use and function of the mandibular gland secretions in *Atta* workers. Of course, it would also be of interest to investigate age-specific changes of mandibular gland secretions in the worker subcastes and to analyze worker reactions to these components in different locational and behavioral contexts.[100]

Of central significance for the operation of the superorganism is the communication between its queen, acting as its reproductive unit, and the workers, making up its somatic units. In her long life-time, spanning more than ten years, the queen of a large *Atta* colony can produce as many as 150 million daughters, of which the vast majority are workers. Every year, several thousand of these females in mature colonies grow up not into workers but into alate queens, each able to mate and found a new colony on her own. In addition, several thousand of the queen's progeny develop every year from unfertilized eggs to become the short-lived males. It is through the young queens and males that the colony reproduces and propagates its genes. Colonies that produce the largest number of such healthy reproductive forms have the best chance of being represented in the next generation. However, to produce such a large crop, the colony needs a huge worker force to secure and retrieve the resources required to rear the energetically quite expensive sexual brood. It is not too much to say that the sole purpose of workers is to put into the world as many royal siblings as possible.

The gigantic *Atta* queen is surrounded by her daughter workers at all times. She is continuously groomed and fed, and she produces an enormous number of eggs. A rough calculation reveals that the mother of a mature colony lays on average about 20 eggs per minute, thus 28,800 per day and 10,512,000 per year. In the presence of a fertile queen, nest workers as a rule lay only deformed trophic eggs, which are then fed to the queen. The production of viable eggs by workers in intact *Atta* colonies would negatively affect colony efficiency and would be a serious handicap in reproductive competition with other mature colonies in a population. Thus, we should expect that nest workers are continuously informed about their queen's presence and fertility. But how is such a queen-to-worker communication in the huge *Atta* colonies possible? We do not yet know, but we can make a reasonable guess on the basis of other information.

The *Atta* queen appears to remain quite stationary in one of the central fungus garden chambers of the expansive nest structure, where she does nothing but eat mostly trophic eggs and produce reproductive eggs. Her eggs are then distributed by workers over the entire fungus garden. (Such dispersion is necessary, because if the eggs remained in the queen's chamber, she would be suffocated by the growing mass.) Might the scattering eggs themselves carry a cue of the queen's presence? It has recently been shown in the monogynous carpenter ant *Camponotus floridanus* that this form of communication does occur: queen-laid eggs are coated with a queen-specific blend of hydrocarbons that serves as a queen fertility signal.[101] Workers respond to the pheromone by refraining from laying viable eggs. The distribution of queen-laid eggs by the workers spreads the queen signal in the colony. It seems very likely that a similar method of queen signal transmission will be found in colonies of *Atta*.

THE ANT-FUNGUS
MUTUALISM

Whenever two kinds of organisms live in close mutualistic symbiosis, as is the case in leaf-cutting ants and their fungus, we should expect communication between the two mutualists. The fungus may signal to its host ants its preference for particular vegetable substrates or the need for a change in diet to maintain nutritional diversity or even the presence of a harmful substrate. To date, only a few studies have examined the possibility of communication between the fungus and the host ants.

It is well established that selection of the leaf material harvested by the leaf-cutting ants is dependent on both the physical and the chemical characteristics of the plant.[102] It is reasonable to suppose, therefore, that if plant material is loaded with secondary compounds harmful to the fungus, the workers will cease harvesting those plants. However, this reaction might not be immediate. Several hours may pass before the foragers completely abandon this food source.[103] Still, once this delayed "rejection," as it is called, sets in for a particular plant material, the ants continue to refuse it for days or even weeks. How, then, is the information that the harvest material is unsuitable for the fungus transmitted to the foragers?

In laboratory experiments with *Atta* and *Acromyrmex* colonies, P. Ridley and his collaborators demonstrated that the ants learn to reject plant material that contains chemicals harmful to the fungus. Although the foragers initially carried baits containing orange peel laced with cycloheximide, a fungicide, into the nest, they eventually stopped collecting the bait, and the rejection was maintained for many weeks. The test colonies also rejected orange peel not contaminated with the fungicide substance. The researchers hypothesized that if the substrate causes toxic effects on the fungus, the fungus will produce a chemical signal that acts as a negative reinforcement to the ant servicing that fungus garden.[104] In a follow-up study, investigators attempted to trace the pathway of this putative fungal signal.[105] Their results suggest that a signal produced by the fungus does not affect the foragers directly; instead, nonforager workers have to have contact with the fungus in order for the rejection to occur. The results thus suggest that the information is transferred from the smaller fungus garden workers to the larger forager workers.

The hypothetical chemical signal produced by stressed fungal tissue has yet to be characterized. Meanwhile, R. D. North and his collaborators have proposed an alternative hypothesis: rejection occurs when ants detect fungal breakdown products from unhealthy or dead fungus. The workers then associate dead fungus with "orange flavor" and consequently reject all substrate containing orange.[106] At least this much is known: leaf-cutting ants learn to associate odor with food,[107] as indicated by the fact that workers who have experienced contaminated orange peel also then reject uncontaminated orange peel. Further evidence of the existence of associative learning is that leafcutter workers, if exposed in the nest to particular odors by incoming scouts, then tend to seek material with that odor during their own foraging excursions.[108] Still unanswered by the sick-fungus hypothesis is the means by which the garden workers perceive the health of the fungus and the signals by which they transmit this information to the foragers.[109] In a new study tailored to resemble more natural conditions, Hubert Herz and his collaborators manipulated leaf suitability for the fungus by infiltrating the plants with a fungicide (cyclohexidine) not detectable to the ants. The ants' delayed rejection behavior was specific toward the respective fungicide-treated plant species. The rejection began ten hours after treated leaves were carried into the fungus garden, and it continued for at least nine weeks. Rejection was also observed in naive ants after contact with the fungus garden containing treated leaves. However, acceptance resumed after three weeks when ants were "force-fed" on untreated leaves of the previously treated plant species. This shows again that ants get information from the fungus that a particular plant species is not good for the fungus. The ants identify this plant species and thereafter avoid it as harvest material. They will, however, resume harvesting this plant species when the fungus does not exhibit a negative reaction. This species-specific flexible reception of unsuitable substrate may be a mechanism to avoid provisioning the fungus garden with plants containing harmful compounds, as such plants occur in the highly diverse natural habitats of the leafcutter ant colonies.[110]

Yet another kind of fungal signaling exists in leaf-cutting ant colonies: the ants recognize their own symbiotic fungal strain and protect it against competing strains introduced from other colonies.[111] Experiments by Michael Poulsen and Jacobus Boomsma have recently revealed that the mechanism where this discrimination behavior originates is in the fungus.[112] The researchers used fungus gardens from colonies of two sympatric species of Panamanian leaf-cutting ants, *Acromyrmex echinatior* and *Acromyrmex octospinosus*. The clonal fungi of both species belong to the same genetically diverse clade. The compatibility of fungi from different colonies was assessed by inoculating pairs of mycelia 1.5 centimeters apart on an agar medium. After two months, mycelial intercompatibility could be measured on a scale from fully compatible to total rejection. By this means, it was demonstrated that, upon contact, domesticated fungi actively reject mycelial fragments from foreign (even neighboring) colonies. The intensity of the rejection is proportional to the overall genetic differences between the symbionts. Incompatibility compounds were detected in the fungal strains; their chemical structure has yet to be determined.

All fungus-growing ants manure the fungal garden with their own feces. Amazingly, the fungal enzymes that biochemically break down plant material are preserved during passage through the ant gut. After the ant has eaten fungus, the enzymes accumulate with other fecal matter in the rectal bladder of the ants. Fecal droplets containing the recycled enzymes are then deposited on the freshly cut pieces of leaf-bearing mycelial inocula or directly onto older fungal growth. Fecal droplets from ants of a foreign colony cause the same incompatibility effect on mycelial growth as the direct introduction of a foreign fungus. The intensity of rejection by fungi toward fecal droplets from nonresident ants corresponds to the genetic distance between the inoculum and the resident fungus receiving it. Oddly, initial incompatibility is lost and changes to compatibility when ants are forced to feed on an incompatible symbiont for ten days or longer. The ants' new fecal droplets then become incompatible with their original resident fungus. From these striking results, Poulsen and Boomsma concluded that the symbiotic fungus uses its host ants to carry the fungus-specific incompatibility signal to all parts of the vast fungus garden of the leafcutter ant colony. Their results suggest that the ants' manuring practice is the decisive factor that constrains colonies to rearing a single clone of symbiont. Obligate manuring with feces allows the resident fungus to control the genetic identity of new gardens in the nest, causing the removal of unrelated fungi before they contribute to ant feeding, and hence ensuring the production of compatible fecal droplets.[113]

Let us summarize to this point. Because hostile interactions between symbionts reduce overall productivity, the introduction of an alien fungus clone not only harms the resident fungus but also diminishes the growth and productivity of the host ant colony. It is therefore in the interest of both the resident fungus and the host to avoid competing fungal strains. The purity of the resident fungal clone is maintained through the action of the fungal incompatibility compound contained in the ants' fecal droplets.[114]

However, one has to be careful not to overemphasize the role of the "reciprocal antagonism" believed to underlie mutualism between fungus and fungus-growing ants. In fact, the findings by Jon Seal and Walter Tschinkel present an intriguing challenge to this hypothesis. They conducted a series of fungus switch experiments, by replacing the native fungus from nests of *Trachymyrmex septentrionalis* with fungus from the leafcutter ant *Atta texana*. The results are stunning. Harvest preferences were not altered in colonies with the new cultivar, and generally the *Atta* fungus did not affect behavior and fungus tending in workers of *Trachymyrmex,* compared with colonies where the native fungus has been retained. Nor were the production of reproductive individuals and the sex ratio affected by the fungus switch. From these and other findings Seal and Tschinkel conclude "that cooperation and not conflict has been more important in shaping the evolutionary ecology of this mutualism. Although the cultivars were certainly genetically and physiologically distinct, these differences did not account for variation in the production of ant biomass or ant behavior. The emerging picture thus indicates that mutualism should be viewed as a highly integrated superorganism that is more than the sum of its parts."[115]

All this symbiotic webwork beautifully illustrates how much the symbiotic fungus has evolved to be an intricate part of the leafcutter ant superorganism. Neither of the mutualistic partners would be able to exist alone. The ants' division of labor and much of their social behavior are shaped by the details of this symbiotic relationship. In turn, the productivity and clonal propagation of the fungus is entirely dependent on its host ants. Although there might be some evolutionary conflict of interest and some exploitative manipulation for fitness advantages on both sides of the relationship, each kind of organism must be evolutionarily adjusted to the other, or the colony dies.

9

HYGIENE IN THE SYMBIOSIS

Maintaining a high level of vitality and hygienic condition of the fungus gardens is crucial for the host ant colony to survive and reproduce. An adequate level is not easy to accomplish; for the fungus to flourish, the requisite subterranean growth chambers need high humidity and tropical temperatures. The ants keep their gardens clean by an impressive variety of hygienic techniques: they pluck out alien fungi; they inoculate the correct fungal mycelia onto fresh substrate; they fertilize the substrate with fecal droplets that contain incompatibility substances to repel alien strains of the host fungus species; they secrete antibiotics to depress competing fungi and microorganisms; and they produce growth hormones.[116]

In 1970, Ulrich Maschwitz and his collaborators made the pioneering discovery that antibiotic substances are produced in the metapleural glands of *Atta sexdens* workers.[117] These paired glandular structures are located at the side near the distal end of the ant's middle body segment called the mesosoma or alitrunk. They suggested that the compounds play different roles in the purification of the symbiotic fungus culture: phenylacetic acid suppresses bacterial growth; myrmicacin (hydroxydecanoic acid) inhibits the germination of spores of alien fungi; and indoleacetic acid, a plant hormone, stimulates mycelial growth.[118] Recently, a more comprehensive analysis of metapleural gland secretions of *Acromyrmex octospinosus* revealed twenty previously unrecognized compounds.[119] They span the whole range of carboxylic acids, from acetic acid to long-chain fatty acids, in addition to keto acids, alcohols, and lactones.

The metapleural glands of leaf-cutting ant workers are relatively large compared with those of other kinds of ants; interestingly, that is particularly true in the smallest workers.[120] The latter disproportion suggests that the allocation of resources to metapleural gland secretions is most important in the minor workers, who predominantly tend the fungus and care for the brood.

The earlier prevailing assumption that fungus-culturing ants maintain their fungus gardens in completely pure conditions has had to be revised with the later discovery that fungus gardens are often contaminated by bacteria, yeasts, and other kinds of fungi.[121] A more thorough and deeper search for pathogens and parasites in leaf-cutting ant colonies demonstrated that while the ants cannot prevent contamination, they are nonetheless able to hold the growth of the invading microorganisms and foreign fungi to very low levels. It has been suggested that the main countermeasure of the ants against parasitic fungi is to maintain the fungus cultures at an acidic pH of 5, optimal for the symbiotic fungus but detrimental for pathogenic invading fungi.[122] Supporting this hypothesis is the fact that the pH rises to 7 or

8 when the ants are removed, and within a few days, parasitic fungi and bacteria spread rapidly in the symbiotic fungus cultures. For that reason, it has been suggested that one of the main functions of the metapleural gland secretions of *Acromyrmex* and *Atta* workers is to reduce the pH of the leaf material brought into the colony from approximately 7 or 8 to 5. It is an added benefit that each of the acids present in the secretions also has antibiotic properties.[123]

Striking new discoveries have recently been reported concerning the "agricultural pathology" of ant fungus gardens. By extensive isolation of nonmutualistic fungi from the gardens of attine ants, Cameron Currie and his collaborators found specialized garden parasites belonging to the microfungus genus *Escovopsis* (Ascomycota: anamorphic Hypocreales). These parasites are horizontally transmitted between leaf-cutting ant colonies. *Escovopsis* is highly virulent, able to devastate ant gardens and thus doom the entire ant colony (Plate 38). Most remarkably, the genus *Escovopsis* appears to specialize on fungal gardens of attine ants. It has not been isolated from any other habitat, and it is especially prevalent in *Atta* and *Acromyrmex* colonies.

Currie and co-workers explain the success of the parasite with the following argument. The increased prevalence of *Escovopsis* within the more derived attine genera suggests that the long clonal history of the leafcutter fungal cultivars, perhaps as long as 23 million years, makes them more susceptible to losing the "arms race" with parasites. By contrast, lower attines routinely acquire new fungal cultivars from free-living sexual populations, leading to a greater genetic diversity in the fungal mutualist population. This may account for the apparent lower susceptibility to parasitism of the less derived attine lineages.[124]

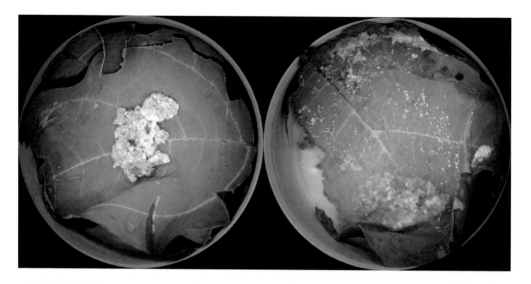

PLATE 38. The parasitic fungus *Escovopsis*. *Upper*: On the left is a piece of intact fungus from the fungus garden of an *Atta* nest. On the right the *Atta* fungal cultivar is overgrown with *Escovopsis* filaments, after the *Atta* fungal cultivar was inoculated with *Escovopsis* spores. (Preparation and photo: Michael Poulson.) *Lower*: Filamentous hyphae of the fungal parasite *Escovopsis* overgrowing a fungus garden of *Atta colombica* in Panama. (Photo: Hubert Herz.)

In opposition to this hypothesis, however, is recently obtained evidence of sexual recombination in fungal symbionts of leafcutter ant taxa. If that phenomenon proves widespread, it suggests that clonality and vertical transmission have not played the critical role in leafcutter symbiotic evolution.[125]

Whatever the cause of the virulence, the question remains, How do the fungus-growing *Atta* and *Acromyrmex* cope with this continuous deadly threat? Obviously, the successful maintenance of a healthy fungus garden involves a constant struggle to control the *Escovopsis* incursions. Some deterring effect might come from the metapleural gland secretions. But the main weapon against *Escovopsis* appears to be a third mutualist associated with attine ants, an actinomycetous filamentous bacterium of the genus *Pseudonocardia*.[126] This symbiont produces antibiotics that strongly suppress the growth of *Escovopsis*.[127] A recent analysis of such a system at the molecular level revealed that the actinobacterium associated with the fungus-growing ant *Apterostigma dentigerum* produces a substance, called dentigerumycin, that selectively inhibits the parasitic fungus *Escovopsis*.[128] *Pseudonocardia* bacteria are true, evolved mutualists; they inhabit regions of the ants' cuticle that appear to be specific to the ant genus. In *Acromyrmex*, for example, they are housed on the laterocervical plates of the propleura (Figure 6). In this region, the *Acromyrmex* ants possess morphological modifications, such as crypts lined with integumental protrusions in the form of tubercles. Numerous exocrine gland cells connect to the tubercles through cuticular channels. The mutualistic filamentous bacteria are housed inside the crypts. Such bacterium-harboring structures have been found so far only in fungus-growing ants. However, the form and the location of these structures are highly variable within the phylogeny of attine ants,[129] and even in *Acromyrmex* workers the actinomycetes frequently reside also on head and thorax, making the

FIGURE 6. An actinomycete filamentous bacterium of the genus *Pseudonocardia* is a symbiont of leafcutter ants. It produces antibiotics that strongly suppress the growth of parasitic fungi. In the ant *Acromyrmex,* this symbiont is housed in the frontal region of the thorax (the laterocervical plates of the propleura), where it lives in special bacteria-harboring structures. It can spread, however, beyond this structure and sometimes covers large parts of the body. (Illustration by Margaret Nelson.)

ant look as if covered with white dust (Plate 39). It is likely that the secretions of the glandular crypts help maintain the actinomycete cultures.

Still more amazing is the way the fungus garden workers prevent the invasion and spread of microbial parasites. When grooming the fungus (Plate 40), the ants collect debris and pathogenic spores, such as those of *Escovopsis*, and load them into their infrabuccal pocket, the content of which is later expelled in debris chambers or midden heaps outside the nest. The infrabuccal pocket is located in the oral cavity of all ant species and serves as

PLATE 39. *Acromyrmex* workers carry the whitish symbiotic actinomycete *Pseudonocardia* on the cuticle of their bodies. Sometimes the ant's entire body surface is covered by a thick filamentous growth of the actinobacterium. The symbiotic bacteria produce specific antibiotic substances against the parasitic fungus *Escovopsis*. (*Upper* photos: Christian Ziegler. *Lower* photo: Michael Poulsen.)

PLATE 40. Garden worker ants clean the fungus garden of *Atta cephalotes* of foreign fungal spores, debris, and parasitic fungal tissues. (Photo: Alex Wild.)

a food filtering device. In the fungus-growing ants this structure is used as a specialized sterilization "chamber," killing spores of garden parasites. Presumably, actinomycetous bacteria inside the infrabuccal pocket produce the antibiotics that inhibit the spores' development.[130]

The mutualist bacteria are transmitted vertically (from parent to offspring colonies) on the body of the founding queen, in the same fashion as the symbiotic fungus. The bacteria are not just adapted to fight the parasitic fungus. They also promote the growth of the symbiotic fungus in vitro. In extreme cases of infestation, the ant colony may be forced to escape from *Escovopsis* by nest emigration, carrying their bacteria with them, to continue the struggle in a new location.[131]

We also have to consider the rich community of microorganisms inside the leafcutter nests. In one study, nineteen species of microorganisms associated with leafcutter ants were characterized, of which one strain of *Streptomyces* proved highly potent in inhibiting *Escovopsis* growth.[132] From the compounds isolated from *Streptomyces,* the substance candicidin turned out to be a highly active chemical agent against *Escovopsis* in laboratory tests, and at least one of the microbial isolates from each of the *Acromyrmex* species produced candicidin.[133] Thus, this might be another chemical mace against microbial parasites in the fungus garden. In fact, actinomycete bacteria were also detected on the cuticle of ant species that do not grow fungi. The presence of fungicide-producing bacteria is obviously beneficial for soil-dwelling insects exposed to multiple risks of infection. The diverse assemblages of actinomycetes that produce a variety of fungicides could provide the ants additional protection against fungal infection. Interestingly, neither the compound isolated from *Pseudonocardia* nor the one isolated from *Streptomyces* affects growth of the leafcutters' symbiotic fungus.

Of course, in speaking of symbiosis one must ask whether the association of microbes with leafcutters and their fungus on the one hand, and the parasitic specialist *Escovopsis* on the other, have a deep coevolutionary history or instead are incidentally assembled in their particular nest habitat. For example, some microbes might have resided on leaf fragments carried by leafcutters into the nest. These questions await resolution in future research.

There is, however, diverse and very convincing evidence that

Pseudonocardia is part of a truly coevolved symbiotic system in fungus-growing ants.[134] The uniqueness and tight fit in the relationship between attine ants and *Pseudonocardia* bespeak an ancient origin of the mutualism. Cameron Currie and his colleagues conclude overall, "Although the ant-fungus mutualism is often regarded as one of the most fascinating examples of highly evolved symbiosis, it is now clear that its complexity has been greatly underestimated. The attine symbiosis appears to be a co-evolutionary 'arms race' between the garden parasite, *Escovopsis*, on the one hand, and the tripartite association amongst the actinomycete, the ant host, and the fungal mutualist on the other" (Figure 7).[135]

Astonishingly, this is not yet the complete story of these highly complex symbiotic systems in leafcutter colonies. A fourth player has recently been discovered. It is a black yeast that eats *Pseudonocardia* on the ants' cuticle and thereby negatively affects the efficiency of fighting the parasitic *Escovopsis* fungus with antibiotic compounds derived from *Pseudonocardia*. Indeed, ants infected with black yeasts are significantly less effective at defending their fungus garden from *Escovopsis* invasion.[136]

And finally, a fifth player in this symbiotic network has recently been detected. Scientists associated with Cameron Currie were able to isolate nitrogen-fixing bacteria from fungus gardens of eighty leafcutter ant colonies. Nitrogen-fixing bacteria use nitrogen from the atmosphere and convert it to ammonia. This is an essential life process because ammonia is the basic building block for amino acids and proteins. Relatively few plants, mainly in the leguminous family Fabaceae, have the capacity to fix atmospheric nitrogen. However, they can do this

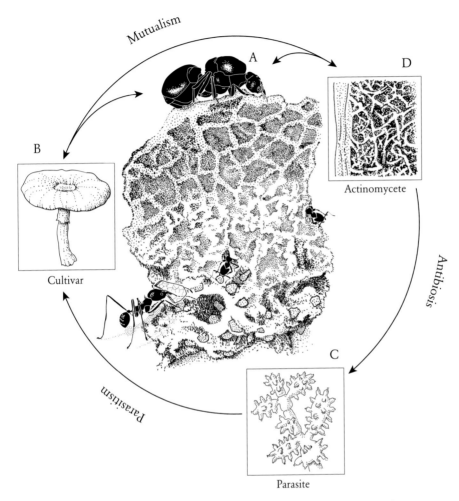

Mutualism

Antibiosis

Parasitism

Cultivar

Actinomycete

Parasite

FIGURE 7. The quadripartite symbiosis of leafcutter ants. A: The queen is the reproductive unit of the leafcutter colony. B: Mushroom habitus of the free-living leucocoprineous fungi. C: The parasitic microfungus *Escovopsis*. D: The filamentous actinomycete *Pseudonocardia*, which grows on the cuticle of the ants and produces antibiotics that suppress the growth of *Escovopsis*. The arrows indicate interacting components. (Redrawn by Margaret Nelson, based on an original illustration by Cara Gibson in C. R. Currie, "A community of ants, fungi, and bacteria: a multilateral approach to studying symbiosis," *Annual Review of Microbiology* 55: 357–380 [2001].)

only with the aid of symbiotic bacteria, which live in the plants' root systems. Currie and co-workers demonstrated that atmospheric nitrogen-fixing bacteria (most likely of the genus *Klebsiella*) live in the ants' fungus gardens and facilitate the cultivation of the leafcutters' symbiotic fungus.[137]

1 0

WASTE MANAGEMENT

The depleted substrate left by the fungus forms a tremendous residue. Most *Atta* species build special refuse chambers in their nests to receive this waste, but *Atta colombica* has another solution: its colonies dispose of the material outside the nest. The refuse is loaded with secondary plant compounds and possibly parasitic fungal mycelia and other pathogens. The ants exhibit a strong avoidance of the waste once it is removed. Among many local people in Latin America, it has long been known that refuse from *Atta* nests can be used as a powerful repellent against ants. Experiments have shown that the *Atta* waste material scattered around young plants protects them from *Atta* herbivory.[138]

Carl Anderson and Francis Ratnieks report fascinating experiments that demonstrate how *Atta colombica* colonies usually manage their external waste removal by a form of task partitioning.[139] The refuse is taken from the nest and deposited on a cache along the trail to the dump (Plate 41). Other workers collect material from the cache and carry it to the main pile. The adaptive value of partitioning the task of waste removal may be the reduction of the spread of disease

PLATE 41. Refuse mound of *Atta colombica*. Unlike most other *Atta* species, which have refuse chambers inside the nest, *Atta colombica* disposes of its refuse outside the nest. (Photo: Hubert Herz.)

and parasites into the colony by segregating the garbage collectors inside the nest from the dump managers outside.[140] Ants exposed to waste material die at a higher rate, and waste is often infected by the fungal parasite *Escovopsis*. Waste management is mainly performed by older workers, who are destined soon to die anyway.[141] The proneness of older workers to risk their lives is obviously an adaptive trait with respect to colony-level efficiency. This phenomenon of older workers taking greater risks is true for many ant species and in different contexts.[142]

Occasionally, as we have noted, a fungus garden is struck by a massive invasion of *Escovopsis* or other pathogen, and the colony is forced to abandon its nest and fungus garden and emigrate to a new nest site. The colony then has to acquire a new fungal cultivar. Genetic data suggest that transfer of fungus from one colony to another can occur, and for the lower attine genus *Cyphomyrmex*, this phenomenon has been demonstrated experimentally.[143] In the laboratory, we have repeatedly "healed" declining gardens of *Atta cephalotes* colonies by introducing pieces of healthy gardens cultivated by other colonies. In this context, the reports of raids of incipient colonies by large colonies of *Atta sexdens rubropilosa,* with the transfer of brood and fungus material, are of special interest.[144] These observations were made with colonies cultured in the laboratory, and no such documentation exists as yet for *Atta* in the field. However, raids have been observed in nature between incipient colonies of *Acromyrmex versicolor.*[145] It seems likely that the stealing or usurpation of fungus gardens by older attine colonies whose own garden has been devastated does occur as a natural response to the loss of fungus gardens.[146]

11

AGROPREDATORS AND
AGROPARASITES

The symbiotic fungus of attine colonies obviously is an attractive resource, not only for parasitic fungi that require the garden substrate and conspecific ant colonies that have lost their own fungus garden, but also for other ant species. Michiel Dijkstra and Jacobus Boomsma have described the predatory raids conducted by the ponerine ant *Gnamptogenys hartmanni* in Panama. *Gnamptogenys* scouts that have discovered a fungus-growing ant nest of *Trachymyrmex* or *Sericomyrmex* species recruit raiding columns of nestmates that attack and usurp the nest of the assaulted colony. There is hardly any defense; the workers of the raided colony flee their nest in panic. The raiders' colony then moves into the captured nest, where its workers and larvae consume the fungus and host brood. After this resource is depleted, the colony conquers the nest of another fungus-growing ant colony.[147]

This raiding behavior closely resembles that discovered in the myrmicine genus *Megalomyrmex* (Plates 42 and 43).[148] These agropredators

PLATE 42. *Upper*: Queen and workers of the agroparasite *Megalomyrmex symmetochus* collected in Panama. *Lower*: Workers of the same species, which is a social parasite of the fungus-growing ant *Trachymyrmex*. They eat the fungus of their host ants. (Photos: Alex Wild.)

PLATE 43. The parasitic *Megalomyrmex symmetochus* workers and their host ant *Trachymyrmex* (the larger individuals). (Photos: Alex Wild.)

raid nests of the attine ant *Cyphomyrmex longiscapus,* and they consume the cultivated fungus and the attine brood after all the resident ants have been killed or expelled. Raiding *Megalomyrmex* species may represent an early evolutionary grade in the phylogenetic pathway toward a trophic and social parasitic cohabitation and usage of the fungus garden with the host ants, an adaptation described for *Megalomyrmex symmetochus.*[149]

Surprisingly, no agropredators are yet known to attack the leaf-cutting genera, *Atta* and *Acromyrmex.* It might be that their nests are too large to be usurped, and their elaborate worker subcaste systems with specialized defenders make them resistant to such predation. However, at least two kinds of social parasitic ants live in *Acromyrmex* colonies. *Pseudoatta argentina,* which parasitizes colonies of *Acromyrmex lundi,* is a highly derived social parasite in which the worker caste has been lost. Another social parasite, *Acromyrmex insinuator,* apparently represents a less derived evolutionary grade of social parasitism. It retains a worker caste, and its morphology still closely resembles that of its host species, *Acromyrmex octospinosus.* These social parasites coexist intimately with their host ants and eat their fungus, but do not participate in the culturing efforts.[150] Although obviously an economic burden on the leaf-cutting ant colony, they do not go so far as to devastate the fungus garden.

12

LEAFCUTTER NESTS

The enormous number of worker ants and the huge fungus gardens of an *Atta* colony require a colossal nest capacity (Plate 44). One typical *Atta sexdens* nest, more than six years old, contained 1,920 chambers, of which 238 were occupied by fungus gardens and ants. The loose soil that had been brought out and piled on the ground by the ants during the excavation of their nest weighed approximately 40,000 kilograms (40 tons). Although nests of different *Atta* species have been excavated and reconstructed on paper by several authors,[151] the recent, quantitatively detailed work by Luiz Forti and his team in Brazil has delivered a breakthrough in our understanding of the megalopolis architecture of *Atta* colonies.[152]

The nest mound of the mature *Atta laevigata* colonies measured by these researchers varied from 26.1 to 67.2 square meters. In addition to a careful step-by-step excavation, the team perfected a method for making a mold of the nest interior. To obtain the mold, they poured liquid cement into the nest entrance. For one large nest, a mixture of 6,300 kilograms (6.3 tons) of cement and 8,200 liters of water was required, enough for a small human dwelling. After two to three weeks,

PLATE 44. The mature nests of *Atta* species are gigantic. Shown here is a nest of *Atta vollenweideri* in Argentina. (Photo: Flavio Roces.)

the preserved nest structure was carefully excavated (Plates 45 to 48).[153] The number of nest chambers in the sample made by the Forti team ranged from 1,149 (smaller mature colony) to 7,864 (largest colony), both reaching as deep as seven to eight meters underground. Most of the chambers were located at a depth of one to three meters. In the very large nests, about 30 percent of the chambers were found below four meters, although several were empty. Some chambers contained declining fungus cultures, but many others housed flourishing fungus gardens complete with brood and ants. In addition, a number of chambers were filled with plant debris and degraded fungal material. Extensive subterranean foraging tunnels led into a central area of the highest concentration of fungus garden chambers. Smaller tunnels branched off from the main channels, and still smaller ramifications connected directly to individual fungus garden chambers. Most of the garden chambers had only one such small tunnel, with an opening located in the middle or near its base. The volume of the largest chambers ranged from about 25 to 51 liters, and that of the smallest ones from 0.03 to 0.06 liter.

The nests of all species of *Atta* possess a comparatively complex architecture, comprising many tunnels and chambers of variable sizes and shapes.[154] A feature common to nests of at least *Atta laevigata, Atta sexdens rubropilosa, Atta vollenweideri,* and *Atta bisphaerica* is the location of fungus garden chambers mainly below the loose soil mound and down to about three meters deep. The Brazilian researchers have suggested the following architectural rationale: "Probably the accumulation of loose soil over the fungus chambers has the purpose of thermal insulation, since *Atta laevigata* often build their nest in open habitats with the fungus chambers rather close to the surface. From the localization of the fungus chambers we must conclude that the upper 3 m of soil offer the best microclimatic conditions for fungus growth."

An *Atta* nest thus houses the huge fungal body together with millions of workers and immature ants in a distributed but interconnected network of nest chambers. As the enormous biomass metabolizes, it produces large quantities of carbon dioxide, which can be fatal to the ant colony if the concentration becomes too high. *Atta* workers are equipped with very sensitive carbon dioxide receptors on their antennae, enabling them to measure carbon dioxide concentrations.[155]

Concentrations of carbon dioxide inside nests of *Atta vollenweideri* vary according to the size of the nests, to differences in the effectiveness of nest ventilation, and finally to differences

PLATE 45. A mature nest of *Atta laevigata* in Brazil was excavated after six tons of cement and 8,000 liters of water had been poured into the nest to preserve its structure. (Photo: Wolfgang Thaler.)

between small and large colonies. Small colonies tend to close their nest entrances during rain to protect the fungus garden from flooding. In such situations, carbon dioxide concentrations increase rapidly, and colony respiration rates are reduced. It appears that the ants' respiration remains unchanged, but the respiration of the symbiotic fungus is lowered. This subsidence, of course, negatively affects the growth rate of the fungus, and ultimately that of the colony, because the fungus is the main food source of the larvae. Young growing colonies are thus confronted with a trade-off: minimizing the risk of being flooded and drowned versus providing adequate gas exchange inside the nests.[156]

PLATE 46. Portions of the cement-filled subterranean tunnels, ducts, and fungus chambers of the *Atta laevigata* nest. (Photo: Wolfgang Thaler.)

In nests of mature colonies, on the other hand, provided with many nest openings and with deep chambers, gas exchange occurs without interruption, albeit with variable intensity. The openings in the central area of the nest mound are often shaped like turrets (see Plate 44). Investigators have found a markedly negative correlation between wind velocity above the nest mound and carbon dioxide concentration inside the nest, with the wind on the surface most likely inducing an increased outflow of carbon dioxide–laden air. The decaying organic material in the refuse chambers often raises the ambient temperature within the nest. If the outside temperature is lower, warmer carbon dioxide–laden air from the nest chambers rises and flows

PLATE 47. *Left*: The ball-shaped structures connected by tunnels and ducts are preserved fungus chambers. *Right*: A live fungus chamber. (Photos: Wolfgang Thaler.)

out of the turrets, with cooler fresh air pulled back into the nest through other entrance tubes.[157]

It follows that both passive ventilation due to external wind velocity and thermal convection likely drive the gas exchange system in *Atta vollenweideri* nests and probably also in nests of other *Atta* species. *Atta* nests located in open grasslands are more exposed to strong winds than nests located in forest habitats, such as those of *Atta cephalotes*. In the latter species, thermal convection is probably more important.

Leaf-cutting ants do not depend entirely on nest architecture

PLATE 48. Cast of a subterranean tunnel in a nest of *Atta laevigata*. The subterranean highway leads the ants to the fungus chambers left and right of the tunnel. (Photo: Wolfgang Thaler.)

to regulate nest ventilation. The gardening workers are also able to sense differences in relative humidity and establish fungus gardens in chambers with the highest humidity. When the chambers start to dry out, workers tear up and relocate the garden into chambers with higher humidity.[158] In fact, Martin Bollazzi and Flavio Roces report fascinating experiments with the leaf-cutting ant *Acromyrmex ambiguus.* In the laboratory, a colony was exposed to either dry or humid air flowing through the nest chambers. Circulation of dry air triggered an increase in building activity, and tunnels through which the dry air entered the nest were plugged at the inflow, but much less at the outflow sections. Inflow of humid air did not elicit much, if any, plugging behavior. The direction of airflow served as an environmental cue for spatial guidance of building activities, and control of the nest climate may be a major determinant in the design of nest structures.[159]

In addition, soil temperature appears to be an important cue that affects digging behavior.[160] Indeed, it has been demonstrated that workers of the leafcutters *Atta vollenweideri* can be trained to use thermal radiation as an orientation cue, and special temperature sensitive sensilla have been found on the ants' antennae.[161]

13

TRAILS AND TRUNK ROUTES

Y et another architectural feature of *Atta laevigata* nests is the extensive horizontal foraging tunnels, constructed about 40 to 50 centimeters below the ground surface. In cross section, they are elliptical, measuring 4 to 48 centimeters in width and 2 to 6 centimeters in height. In larger nests, the foraging tunnels are wider, but not necessarily higher.[162] These tunnels can extend to 90 meters.[163] They channel the masses of foragers arriving on the aboveground trunk routes, which in turn connect the nest and harvesting areas, sometimes extending more than 250 meters from the nest. Because they are static and long-lived, these trunk routes are considered part of the nest architecture. In most cases, they are deeply engraved into the ground and conspicuous even to the most casual observer. They are retained for months or years. Even when abandoned for a time, they are commonly used again by the ants. Serving as the superhighways of the *Atta* colonies, they are continuously cleaned of invading vegetation and other obstacles by "road workers" (see Plate 33). The trunk route system enhances foraging efficiency by increasing foraging speed four- to tenfold compared with that on uncleared ground.[164] They are, moreover, part of the defended territories that serve to protect the colony's resources from competitors.[165]

Trail construction and maintenance add a considerable amount to the overall energy investment for resource acquisition. However, as documented in a study of *Atta colombica,* the energy expenditure is small relative to the energy yielded by their use, and their cost does not constrain their construction.[166] In addition, most *Atta* species appear to be trail centered. That is, their search for high-quality resources is restricted to the vicinity of the trunk routes.

In one study conducted by Rainer Wirth and his colleagues in Panama, the trunk trail systems of *Atta colombica* colonies were monitored continuously over a period of a full year. The data obtained were then used for an assessment of the true size of the foraging area. One representative nest had four major trunk routes, each of which opened out into numerous branches. The foraging areas of the four routes were 2,712, 2,597, 2,640, and 2,409 square meters. The total of the foraging area visited was thus estimated at an impressive 1.03 hectares (Figure 8).[167] This happens to be approximately the same "ecological footprint," or average amount of land utilized to sustain one person, in developing countries.

Comparative studies of the nest architecture of *Atta laevigata, Atta bisphaerica, Atta vollenweideri, Atta capiguara, Atta sexdens,* and *Atta texana* reveal much similarity but also species-specific differences.[168] Nest structures are the product of innate collective behavior. They are, to use a metaphor coined by Richard Dawkins, the "extended phenotype" of each kind of superorganism in turn. No less than the anatomy and physiology of the ants themselves,

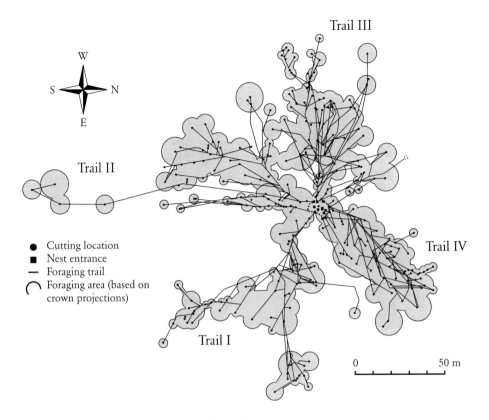

FIGURE 8. The trail system of one colony of the leaf-cutting ant *Atta colombica* covered an entire hectare, as illustrated here. The estimates of foraging area (gray) are based on regression estimates of crown projection area. (From R.Wirth, H. Herz, R. J. Reyel, W. Beyschlag, and B. Hölldobler, *Herbivory of Leaf-Cutting Ants: A Case Study on* Atta colombica *in the Tropical Rainforest of Panama* [New York: Springer-Verlag, 2003].)

architectural features of the nests are shaped by natural selection. They are an exemplary illustration of natural selection acting at the level of the entire society. Like any other group-level units of biological organization, the nests are, to quote Thomas D. Seeley, "elegant devices that nature has evolved for integrating thousands of insects into a higher-order entity, one whose abilities far transcend those of the individual."[169]

A century after William Morton Wheeler's publication of this idea of an insect colony as a superorganism, scientists have revived the superorganism concept with an emphasis on colony-level adaptive demography and organization based on division of labor, visualizing the colony as a self-organized entity and the target of social selection. In particular, David S. Wilson and Elliott Sober have argued that insect colonies can be true superorganisms that are targeted by selection, provided colonies have differential group fitness and the variation in group fitness is caused by heritable variation. In addition, no reproductive competition within groups should exist; or at the very least, in their view, intragroup competition should be conspicuously lower than intergroup competition.[170]

If we were to accept this concept of a superorganism, with its codicil of little or no reproductive competition among nestmates, many of the poneromorph and myrmeciine societies that have been described in the literature[171] might not be considered true superorganisms, because intracolony reproductive competition is indeed conspicuously common. We share the view now generally held by most researchers that insect societies are dynamic, self-organized systems of different complexity. Nevertheless, the thousands of social insect species display among themselves almost every conceivable grade in the division of labor, from little more than competition among nestmates for reproductive status to highly complex systems of specialized subcastes. The level of this gradient at which the colony can be called a superorganism is subjective; it may be at the origin of eusociality (preferred by E.O.W.) or at a high level, beyond the "point of no return," in which within-colony competition for reproductive status is greatly reduced or absent, and between colony competition is rampant (preferred by B.H.).

But whatever criteria may be adopted, there can be little doubt that the gigantic colonies of the *Atta* leafcutters, with their interlocking symbiont communities and extreme complexity and mechanisms of cohesiveness, deserve special attention as the greatest superorganisms on Earth discovered to the present time.

ACKNOWLEDGMENTS

This book is based on one chapter of our recent book *The Superorganism* (2009), substantially expanded. We have written this book because we want to highlight the remarkable societies of the leafcutter ants, the "ultimate superorganisms," and to extend and update the fascinating discoveries made by an international assembly of scientists. Several of them have discussed their research with us, provided literature or still unpublished information, or read and commented on earlier drafts of the manuscript. In particular, we thank Jacobus (Koos) Boomsma, Cameron Currie, Hubert Herz, Christoph Kleineidam, Christina Kelber, Ulrich Mueller, Flavio Roces, Wolfgang Roessler, Ted Schultz, Thomas Seeley, and Rainer Wirth. We made special efforts to appropriately illustrate this book. For excellent photographs providing a striking view of the marvelous agricultural world of the fungus-growing and leaf-cutting ants, we thank Helga Heilmann, Hubert Herz, Manfred Kaib, Michael Poulsen, Flavio Roces, Wolfgang Thaler, Alex Wild, and Christian Ziegler. We also thank Jacob Sahertian for his expert advice during the design of the color plates.

We are most grateful to Kathleen M. Horton. Her superb library research, manuscript production, and editorial advice were most valuable. We also benefited from the encouragement and suggestions provided by our editor, Robert Weil.

The personal research of Bert Hölldobler reported in this book was supported by grants from the German Science Foundation, the National Science Foundation (USA), and Arizona State University. That of Edward O. Wilson was supported by the National Science Foundation (USA).

GLOSSARY

AGGREGATION A group of individuals, composed of more than just a mated pair or a family, that have gathered in the same place but do not construct nests or rear offspring in a cooperative manner (as opposed to a colony). *See* Colony.

AGROPARASITE A social parasite of an attine colony.

AGROPREDATOR A species that feeds on the fungus garden of attine ants.

ALTRUISM In evolutionary biology, behavior that lowers the genetic fitness of the altruist and increases the genetic fitness of others. *See* Genetic fitness.

ARTHROPOD Any member of the phylum Arthropoda, such as a crustacean, spider, millipede, centipede, or insect.

ATTINE ANTS Ants belonging to the taxonomic tribe Attini, all of which are fungus growers.

ATTINI (adjective, attine) The taxonomic tribe of ants that cultivate fungus for food. Two genera, the leaf-cutters *Acromyrmex* and *Atta,* cut fragments of fresh leaves and other vegetation and process the fragments to serve as a bed for fungal growth.

BIOMASS The dry weight of a set of plants, animals, or microorganisms. The set is chosen for convenience; it can be, for example, a colony of insects, a population of wolves, or an entire forest.

BROOD The immature members of a colony collectively, including eggs, nymphs, larvae, and pupae. Eggs and pupae are sometimes not considered members of the society, but they are still referred to as part of the brood.

CASTE Broadly defined, as in ergonomic theory, any set of individuals of a particular morphological type or age-group, or both, that performs specialized labor in the colony. More narrowly defined, any set of individuals in a given colony that is both morphologically distinct and specialized in behavior.

CLADE A species or set of species representing a distinct branch in a phylogenetic tree and hence of single common ancestry.

CLONE A population of individuals all derived asexually from the same single parent.

COLONY A group of individuals, other than a single mated pair, that constructs nests or rears offspring in a cooperative manner (as opposed to an aggregation). *See* Aggregation.

COLONY ODOR The odor found on the bodies of social insects that is peculiar to a given colony. By smelling the colony odor of another member of the same species, an insect is able to determine whether it is a nestmate or not. *See* Nest odor.

COMMUNICATION Action on the part of one organism (or cell) that alters the probability pattern of behavior in another organism (or cell). Communication can be "manipulative," altering the behavior of the receiver to the advantage of the sender of the signal, or it can be to the advantage of both sender and receiver. The latter mode is called reciprocal communication and is most often observed in the social insects.

ECLOSION Emergence of the adult (imago) from the pupa; less commonly, the hatching of an egg.

ECOLOGY The scientific study of the interaction of organisms with their environment, including both the physical environment and the other organisms that live in it.

ENTOMOLOGY The scientific study of insects.

EUSOCIAL As it pertains to a group of individuals, displaying all of the following three traits: cooperation in caring for the young; reproductive division of labor, with more or less sterile individuals working on behalf of individuals engaged in reproduction; and overlap of at least two generations of life stages capable of contributing to colony labor. This is the formal equivalent of the expressions "advanced social" or "higher social," which are commonly used but with less exact meaning.

EVOLUTION Any genetic change in organisms from generation to generation; or, more strictly, any change in gene frequencies within populations from generation to generation.

EXOCRINE GLAND Any gland, such as the salivary gland, that secretes to the outside of the body or into the alimentary tract. Exocrine glands are the most common source of pheromones, the chemical substances used in communication by most kinds of animals.

FILAMENTOUS Shaped like a filament or thread.

GASTER A special term occasionally applied to the metasoma, or terminal major body part, of ants and other aculeate hymenopterans.

GENETIC FITNESS The contribution to the next generation of one genotype in a population relative to the contributions of other genotypes. By definition, the process of natural selection eventually leads to the prevalence of those genotypes with the highest fitnesses.

GENUS (plural, genera) A group of related, similar species. Examples include *Apis* (the four or more species of honeybees) and *Canis* (wolves, domestic dogs, and their close relatives).

GROOMING The licking of the body surfaces of nestmates. Self-grooming also occurs in ants, whereby individuals clean their own bodies both by licking and by stroking with the legs.

HOLOMETABOLOUS Undergoing a complete metamorphosis during development, with distinct larval, pupal, and adult stages. The hymenopterans, for example, are holometabolous.

HYMENOPTERAN Pertaining to the insect order Hymenoptera; also, a member of the order, such as a wasp, bee, or ant.

HYPHA (plural, hyphae) The threadlike growth of a fungus.

INSECT SOCIETY In the strict sense, a colony of eusocial insects (ants, termites, eusocial wasps, or eusocial bees). In the broad sense adopted in this book, any group of presocial or eusocial insects.

LARVA An immature stage that is radically different in form from the adult; characteristic of the holometabolous insects, including the hymenopterans. For the termites, the term is used in a special sense to designate an immature individual without any external trace of wing buds or soldier characteristics.

LIFE CYCLE The entire span of the life of an organism (or of a society), from the moment it originates to the time it reproduces.

MAJOR WORKER A member of the largest worker subcaste, especially in ants. In ants, the subcaste is usually specialized for defense, and adults belonging to this subcaste are often referred to as soldiers. *See also* Media worker; Minor worker.

MANDIBLES The laterally moving jaws of the ants.

MATING FLIGHT *See* Nuptial flight.

MEDIA WORKER In polymorphic ant species containing three or more worker subcastes, an individual belonging to the medium-sized subcaste(s). *See also* Minor worker; Major worker.

MINOR WORKER A member of the smallest worker subcaste, especially in ants; also called a minima. *See also* Nanitic worker; Media worker; Major worker.

MUTUALISM A tight biological relationship between two or more species (symbiosis) of the kind that benefits all of the partners.

NANITIC WORKER A worker of extremely small size, a type usually limited to the first generation of workers produced by a nest-founding queen.

NATURAL SELECTION The differential contribution of offspring to the next generation by individuals of different genetic types but belonging to the same population. Individuals with genetical encoded traits that render them better adapted to environmental conditions will reproduce better and thus will be favored by natural selection. This is the basic mechanism proposed by Charles Darwin and is generally regarded today as the main guiding force in evolution.

NEST ODOR The distinctive odor of a nest, by which its inhabitants are able to distinguish the nest from those belonging to other colonies or at least from the surrounding environment. In some cases, the insects (for example, honeybees and some ants) can orient toward the nest by means of the odor. The nest odor may be the same as the colony odor in some cases. The nest odor of honeybees is often referred to as the hive aura or hive odor. *See also* Colony odor.

NUPTIAL FLIGHT The mating flight of the winged queens and males.

ODOR TRAIL A chemical trace laid down by one insect and followed by another. The odorous material is referred to either as the trail pheromone or as the trail substance.

PETIOLE The first segment of the "waist" of aculeate hymenopterans; actually, the second abdominal segment, since the first abdominal segment (propodeum) is fused to the thorax.

PHEROMONE A chemical substance or blend of substances, usually glandular secretions, used in communication within a species. One individual releases the material as a signal, and another individual responds after tasting or smelling it. Primer pheromones alter the physiology of

individuals and prepare them for new behavioral repertoires. Releaser
pheromones evoke responses directly.

PHYLOGENY The evolutionary history of a particular group of organisms; also, the diagram of the "family tree" that shows which species (or
groups of species) gave rise to others.

PUPA The inactive instar of the holometabolous insects (including the
hymenopterans) during which development into the final adult form
is completed.

QUEEN A member of the reproductive caste in semisocial or eusocial species. The existence of a queen caste presupposes the existence also of
a worker caste at some stage of the colony life cycle. In a functional
definition of the reproductive caste, queens may not be morphologically different from workers; such individuals are now called gamer-
gates, meaning mated worker. If one uses a morphological criterion, the
queen caste is defined by its distinctly different anatomy from that of
the worker caste.

RECRUITMENT A special form of assembly by which members of a society are directed to some point in space where work or other collective
actions are required.

RECRUITMENT TRAIL An odor trail laid by a single scout worker and
used to recruit nestmates to a food find, a desirable new nest site, a
breach in the nest wall, or some other place where the assistance of
many workers is needed, as in defense of the territory.

SOCIAL INSECT In the strict and usual sense (for "true," or "advanced,"
social insects), an insect that belongs to a eusocial species—for example,
an ant, a termite, or one of the eusocial wasps or bees, beetles, thrips,
or aphids. In the broad sense, an insect that lives in a cohesive group, in
which the group members interact in a way that binds them together.

SOCIETY A group of individuals belonging to the same species and organized in a cooperative manner. The diagnostic criterion is reciprocal
communication of a cooperative nature extending beyond mere sexual
activity.

SOCIOBIOLOGY The systematic study of the biological basis of all forms
of social behavior.

SOLDIER A member of a worker subcaste specialized for colony defense.

SPECIES The basic lower unit of classification in biological taxonomy,
consisting of a population or series of populations of closely related and

similar organisms. More narrowly defined, a biological species consists of individuals that are capable of interbreeding freely with one another but not with members of other species under natural conditions.

SPERMATHECA The receptacle in a female insect in which the sperm are stored, also called sperm pocket.

STERNITE A ventral sclerite; a portion of the body wall bounded by sutures and located in a ventral position. *See also* Tergite.

STRIDULATION The production of sound or body vibrations by rubbing one part of the body surface against another. Some insect groups (including grasshoppers, crickets, and many ant species) possess special stridulation devices.

SUPER-MAJOR An especially large major worker (soldier, supersoldier), significantly exceeding in size an ordinary major worker.

SUPERORGANISM A society, such as a eusocial insect colony, that possesses features of organization analogous to the physiological properties of single organisms. The eusocial colony, for example, is divided into reproductive castes (analogous to gonads) and worker castes (analogous to somatic tissue); its members may, for example, exchange nutrients and pheromones by trophallaxis and grooming (analogous to the circulatory system). Among the thousands of known social insect species, we can find almost every conceivable grade in the division of labor, from little more than competition among nestmates for reproductive status to highly complex systems of specialized subcastes. The level of this gradient at which the colony can be called a superorganism is subjective; it may be at the origin of eusociality (preferred by E. O. Wilson) or at a higher level, beyond the "point of no return," in which within-colony competition for reproductive status is greatly reduced or absent (preferred by B. Hölldobler).

SYMBIOSIS The intimate, relatively protracted, and dependent relationship of members of one species with those of another. The three principal kinds of symbiosis are commensalism, mutualism, and parasitism.

TERGITE A dorsal sclerite; a portion of the body wall bounded by sutures and located in a dorsal position. *See also* Sternite.

TERRITORY An area occupied more or less exclusively by an animal or group of animals (such as an ant colony) by means of repulsion through overt defense or aggressive advertisement.

TRAIL PHEROMONE A substance laid down in the form of a trail by one animal and followed by another member of the same species.

TROPHIC EGG An egg, usually degenerate in form and inviable, that is fed to other members of the colony.

WORKER A member of the nonreproductive, laboring caste in semisocial and eusocial species. The existence of a worker caste presupposes the existence also of royal (reproductive) castes. In Hymenoptera, particularly in ants and bees, the worker caste can be defined morphologically, emphasizing the distinctly different morphology of workers and queens in most species, or functionally, as in some ponerine ant species that lack a morphological queen caste, and in several wasp species. In termites, the term is used in a more restricted sense to designate individuals in the family Termitidae that completely lack wings and have reduced pterothorax, eyes, and genital apparatus.

REFERENCES

Chapter 1: The Ultimate Superorganisms

1 | T. L. Erwin, "Tropical forests: their richness in Coleoptera and other arthropod species," *Coleopterists Bulletin* 36(1): 74–75 (1982).

2 | E. J. Fittkau and H. Klinge, "On biomass and trophic structure of the central Amazonian rain forest ecosystem," *Biotropica* 5(1): 2–15 (1973).

3 | B. Hölldobler and E. O. Wilson, *The Ants* (Cambridge, MA: Belknap Press of Harvard University Press, 1990).

4 | S. Higashi and K. Yamauchi, "Influence of a supercolonial ant *Formica (Formica) yessensis* Forel on the distribution of other ants in Ishikari Coast," *Japanese Journal of Ecology* 29(3): 257–264 (1979).

5 | U. Maschwitz and H. Hänel, "The migrating herdsman *Dolichoderus (Diabolus) cuspidatus*: an ant with a novel mode of life," *Behavioral Ecology and Sociobiology* 17(2): 171–184 (1985).

6 | B. Hölldobler and E. O. Wilson, *The Superorganism* (New York: W. W. Norton, 2009).

Chapter 2: The Attine Breakthrough

7 | R. Wirth, H. Herz, R. J. Ryel, W. Beyschlag, and B. Hölldobler, *Herbivory of Leaf-Cutting Ants: A Case Study on* Atta colombica *in the Tropical Rainforest of Panama* (New York: Springer-Verlag, 2003).

8 | For an inspiring comparison of the convergent appearances of agriculture in human and ant societies, we refer to T. R. Schultz, U. G. Mueller, C. R. Currie, and S. A. Rehner, "Reciprocal illumination: a comparison of agriculture in humans and in fungus-growing ants," in F. E. Vega and M. Blackwell, eds., *Insect-Fungal Associations: Ecology and Evolution* (New York: Oxford University Press, 2005), pp. 149–190.

9 | B. Hölldobler and E. O. Wilson, *The Ants* (Cambridge, MA: Belknap Press of Harvard University Press, 1990).

10 | T. R. Schultz and R. Meier, "A phylogenetic analysis of the fungus-growing ants (Hymenoptera: Formicidae: Attini) based on morphological characters of the larvae," *Systematic Entomology*

20(4): 337–370 (1995); T. R. Schultz and S. G. Brady, "Major evolutionary transitions in ant agriculture," *Proceedings of the National Academy of Sciences USA* 105(14): 5435–5440 (2008); U. G. Mueller, S. A. Rehner, and T. R. Schultz, "The evolution of agriculture in ants," *Science* 281: 2034–2038 (1998).

11 | I. H. Chapela, S. A. Rehner, T. R. Schultz, and U. G. Mueller, "Evolutionary history of the symbiosis between fungus-growing ants and their fungi," *Science* 266: 1691–1694 (1994); G. Hinkle, J. K. Wetterer, T. R. Schultz, and M. L. Sogin, "Phylogeny of the attine ant fungi based on analysis of small subunit ribosomal RNA gene sequences," *Science* 266: 1695–1697 (1994).

12 | U. G. Mueller, T. R. Schultz, C. R. Currie, R. M. M. Adams, and D. Malloch, "The origin of the attine ant-fungus mutualism," *Quarterly Review of Biology* 76(2): 169–197 (2001).

13 | T. R. Schultz and R. Meier, "A phylogenetic analysis of the fungus-growing ants (Hymenoptera, Formicidae, Attini) based on morphological characters of the larvae," *Systematic Entomology* 20(4): 337–370 (1995).

14 | U. G. Mueller, T. R. Schultz, C. R. Currie, R. M. M. Adams, and D. Malloch, "The origin of the attine ant-fungus mutualism," *Quarterly Review of Biology* 76(2): 169–197 (2001); P. Villesen, U. G. Mueller, T. R. Schultz, R. M. M. Adams, and A. C. Bouck, "Evolution of ant-cultivar specialization and cultivar switching in *Apterostigma* fungus-growing ants," *Evolution* 58(10): 2252–2263 (2004).

15 | J. K. Wetterer, T. R Schultz, and R. Meier, "Phylogeny of fungus-growing ants (tribe Attini) based on mtDNA sequence and morphology," *Molecular Phylogenetics and Evolution* 9(1): 42–47 (1998); S. L. Price, T. Murakami, U. G. Mueller, T. R. Schultz, and C. R. Currie, "Recent findings in the fungus-growing ants: evolution, ecology, and behavior of a complex microbial symbiosis," in T. Kikuchi, N. Azuma, and S. Higashi, eds., *Genes, Behavior and Evolution of Social Insects* (Sapporo, Japan: Hokkaido University Press, 2003), pp. 255–280.

16 | T. R. Schultz and S. G. Brady, "Major evolutionary transitions in ant agriculture," *Proceedings of the National Academy of Sciences USA* 105(14): 5435–5440 (2008).

17 | U. G. Mueller, S. A. Rehner, and T. R. Schultz, "The evolution of agriculture in ants," *Science* 281: 2034–2038 (1998).

18 | I. H. Chapela, S. A. Rehner, T. R. Schultz, and U. G. Mueller, "Evolutionary history of the symbiosis between fungus-growing ants and their fungi," *Science* 266: 1691–1694 (1994).

19 | R. M. M. Adams, U. G. Mueller, A. K. Holloway, A. M. Green, and J. Narozniak, "Garden sharing and garden stealing in fungus-growing ants," *Naturwissenschaften* 87(11): 491–493 (2000).

20 | A. N. M. Bot, S. A. Rehner, and J. J. Boomsma, "Partial incompatibility between ants and symbiotic fungi in two species of *Acromyrmex* leaf-cutting ants," *Evolution* 55(10): 1980–1991 (2001).

21 | A. S. Mikheyev, U. G. Mueller, and P. Abbot, "Cryptic sex and many-to-one co-evolution in the fungus-growing ant symbiosis," *Proceedings of the National Academy of Sciences USA* 103(28): 10702–10706 (2006).

Chapter 3: The Ascent of the Leafcutters

22 | I. R. Leal and P. S. Oliveira, "Foraging ecology of attine ants in a Neotropical savanna: seasonal use of fungal substrate in the cerrado vegetation of Brazil," *Insectes Sociaux* 47(4): 376–382 (2000).

23 | For reviews, see J. M. Cherrett, "History of the leaf-cutting ant problem," in C. S. Lofgren and R. K. Vander Meer, eds., *Fire Ants and Leaf-Cutting Ants: Biology and Management* (Boulder, CO: Westview Press, 1986), pp. 10–17; and B. Hölldobler and E. O. Wilson, *The Ants* (Cambridge, MA: Belknap Press of Harvard University Press, 1990).

24 | R. Wirth, H. Herz, R. J. Ryel, W. Beyschlag, and B. Hölldobler, *Herbivory of Leaf-Cutting Ants: A Case Study on* Atta colombica *in the Tropical Rainforest of Panama* (New York: Springer-Verlag, 2003); H. Herz, W. Beyschlag, and B. Hölldobler, "Assessing herbivory rates of leaf-cutting ant (*Atta colombica*) colonies through short-term refuse deposition counts," *Biotropica* 39(4): 476–481 (2007); H. Herz, W. Beyschlag, and B. Hölldobler, "Herbivory rate of leaf-cutting ants in tropical moist forest in Panama at the population and ecosystem scales," *Biotropica* 39(4): 482–488 (2007).

Chapter 4: Life Cycle of the Leafcutter Ants

25 | W. E. Kerr, "Tendências evolutivas na reprodução dos himenópteros sociais," *Arquivos do Museu Nacional* (Rio de Janeiro) 52: 115–116 (1962).

26 | E. J. Fjerdingstad, J. J. Boomsma, and P. Thorén, "Multiple paternity in the leafcutter ant *Atta colombica*—a microsatellite DNA study," *Heredity* 80(1): 118–126 (1998).

27 | E. J. Fjerdingstad and J. J. Boomsma, "Queen mating frequency and relatedness in young *Atta sexdens* colonies," *Insectes Sociaux* 47(4): 354–356 (2000).

28 | J. J. Boomsma, E. J. Fjerdingstad, and J. Frydenberg, "Multiple paternity, relatedness and genetic diversity in *Acromyrmex* leaf-cutter ants," *Proceedings of the Royal Society of London B* 266: 249–254 (1999).

29 | T. Murakami, S. Higashi, and D. Windsor, "Mating frequency, colony size, polyethism and sex ratio in fungus-growing ants (Attini)," *Behavioral Ecology and Sociobiology* 48(4): 276–284 (2000).

30 | W. D. Hamilton, "Kinship, recognition, disease, and intelligence: constraints of social evolution," in Y. Itô, J. L. Brown, and J. Kikkawa, eds., *Animal Societies: Theories and Facts* (Tokyo: Japan Scientific Societies Press, 1987), pp. 81–102; P. W. Sherman, T. D. Seeley, and H. K. Reeve, "Parasites, pathogens, and polyandry in social Hymenoptera," *American Naturalist* 131(4): 602–610 (1988).

31 | R. M. M. Adams, U. G. Mueller, A. K. Holloway, A. M. Green, and J. Narozniak, "Garden sharing and garden stealing in fungus-growing ants," *Naturwissenschaften* 87(11): 491–493 (2000). The most convincing evidence in support of the "disease resistance hypothesis" was recently published by W. O. H. Hughes and J. J. Boomsma, "Genetic diversity and disease resistance in leaf-cutting ant societies," *Evolution* 58(6): 1251–1260 (2004). This paper also presents a thorough review of these topics.

32 | J. C. Jones, M. R. Myerscough, S. Graham, and B. P. Oldroyd, "Honey bee nest thermoregulation: diversity promotes stability," *Science* 305: 402–404 (2004).

33 | T. D. Seeley and D. R. Tarpy, "Queen promiscuity lowers disease within honeybee colonies," *Proceedings of the Royal Society of London B* 274: 67–72 (2007).

34 | H. R. Mattila and T. D. Seeley, "Genetic diversity in honeybee colonies enhances productivity and fitness," *Science* 317: 362–364 (2007).

35 | W. E. Kerr, "Tendências evolutivas na reprodução dos himenópteros sociais," *Arquivos do Museu Nacional* (Rio de Janeiro) 52: 115–116 (1962).

36 | E. J. Fjerdingstad and J. J. Boomsma, "Multiple mating increases the sperm stores of *Atta colombica* leafcutter ant queens," *Behavioral Ecology and Sociobiology* 42(4): 257–261 (1998). Similar observations were reported for the African honeybee, *Apis mellifera capensis*, in F. B. Kraus, P. Naumann, J. van Draagh, and R. F. A. Moritz, "Sperm limitation and the evolution of extreme polyandry in honeybees (*Apis mellifera* L.)," *Behavioral Ecology and Sociobiology* 55(5): 494–501 (2004).

37 | R. H. Crozier and R. E. Page, "On being the right size: male contributions and multiple mating in social Hymenoptera," *Behavioral Ecology and Sociobiology* 18(2): 105–115 (1985); R. E. Page Jr., "Sperm utilization in social insects," *Annual Review of Entomology* 31: 297–320 (1986). Also suggested is that polyandry reduces the occurrence of diploid males in honeybee colonies; see D. R. Tarpy and R. E. Page Jr., "Sex determination and the evolution of polyandry in honey bees (*Apis mellifera*)," *Behavioral Ecology and Sociobiology* 52(2): 143–150 (2002).

38 | H. G. Fowler, V. Pereira-da-Silva, L. C. Forti, and N. B. Saes, "Population dynamics of leaf-cutting ants: a brief review," in C. S. Lofgren and R. K. Vander Meer, eds., *Fire Ants and Leaf-Cutting Ants: Biology and Management* (Boulder, CO: Westview Press, 1986), pp. 123–145.

39 | M. Autuori, "La fondation des sociétés chez les fourmis champignonnistes du genre 'Atta' (Hym. Formicidae)," in M. Autuori, M.-P. Bénassy, J. Benoit, R. Courrier, Ed.-Ph. Deleurance, M. Fontaine, K. von Frisch, R. Gesell, P.-P. Grassé, J. B. S. Haldane, Mrs. Haldane-Spurway, H. Hediger, M. Klein, O. Koehler, D. Lehrman, K. Lorenz, D. Morris, H. Piéron, C. P. Richter, R. Ruyer, T. C. Schneirla, and G. Viaud, *L'Instinct dans le Comportement des Animaux et de l'Homme* (Paris: Massone et Cie Éditeurs, 1956), pp. 77–104.

40 | B. Hölldobler and E. O. Wilson, *The Ants* (Cambridge, MA: Belknap Press of Harvard University Press, 1990).

41 | M. Bass and J. M. Cherrett, "Fungal hyphae as a source of nutrients for the leaf-cutting ant *Atta sexdens*," *Physiological Entomology* 20(1): 1–6 (1995).

42 | For a review, see J. M. Cherrett, R. J. Powell, and D. J. Stradling, "The mutualism between leaf-cutting ants and their fungus," in N. Wilding, N. M. Collins, P. M. Hammond, and J. F. Webber, eds., *Insect-Fungus Interactions* (New York: Academic Press, 1989), pp. 93–120. Also see U. G. Mueller, T. R. Schultz, C. R. Currie, R. M. M. Adams, and D. Malloch, "The origin of the attine ant-fungus mutualism," *Quarterly Review of Biology* 76(2): 169–197 (2001).

43 | C. Gomes De Siqueira, M. Bacci Jr., F. C. Pagnocca, O. Correa Bueno, and M. J. A. Hebling, "Metabolism of plant polysaccharides by *Leucoagaricus gongylophorus,* the symbiotic fungus of the leaf-cutting ant *Atta sexdens* L.," *Applied and Environmental Microbiology* 64(12): 4820–4822 (1998).

44 | A. B. Abril and E. H. Bucher, "Evidence that the fungus cultured by leaf-cutting ants does not metabolize cellulose," *Ecology Letters* 5(3): 325–328 (2002).

45 | M. Schiøtt, H. H. De Fine Licht, L. Lange, and J. J. Boomsma, "Towards a molecular understanding of symbiont function: identification of a fungal gene for the degradation of xylan in the fungus gardens of leaf-cutting ants," *BMC Microbiology* 8(40): 1–10 (2008).

46 | P. D'Ettorre, P. Mora, V. Dibangou, C. Rouland, and C. Errard, "The role of the symbiotic fungus in the digestive metabolism of two species of fungus-growing ants," *Journal of Comparative Physiology B* 172(2): 169–176 (2002).

47 | R. J. Quinlan and J. M. Cherrett, "The role of fungus in the diet of the leaf-cutting ant *Atta cephalotes* (L.)," *Ecological Entomology* 4(2): 151–160 (1979).

Chapter 5: The *Atta* Caste System

48 | H. G. Fowler, V. Pereira-da-Silva, L. C. Forti, and N. B. Saes, "Population dynamics of leaf-cutting ants: a brief review," in C. S. Lofgren and R. K. Vander Meer, eds., *Fire Ants and Leaf-Cutting Ants: Biology and Management* (Boulder, CO: Westview Press, 1986), pp. 123–145.

49 | J. K. Wetterer, "Nourishment and evolution in fungus-growing ants and their fungi," in J. H. Hunt and C. A. Nalepa, eds., *Nourishment and Evolution in Insect Societies* (Boulder, CO: Westview Press, 1994), pp. 309–328.

50 | E. O. Wilson, "Caste and division of labor in leaf-cutter ants (Hymenoptera: Formicidae: *Atta*), I: The overall pattern in *A. sexdens,*" *Behavioral Ecology and Sociobiology* 7(2): 143–156 (1980); E. O. Wilson, "Caste and division of labor in leaf-cutter ants (Hymenoptera: Formicidae: *Atta*), II: The ergonomic optimization of leaf cutting," *Behavioral Ecology and Sociobiology* 7(2): 157–165 (1980); E. O. Wilson, "Caste and division of labor in leaf-cutter ants (Hymenoptera: Formicidae: *Atta*), III: Ergonomic resiliency in foraging by *A. cephalotes,*" *Behavioral Ecology and Sociobiology* 14(1): 47–54 (1983); E. O. Wilson, "Caste and division of labor in leaf-cutter ants (Hymenoptera: Formicidae: *Atta*), IV: Colony ontogeny of *A. cephalotes,*" *Behavioral Ecology and Sociobiology* 14(1): 55–60 (1983).

51 | I. Ebil-Eibesfeldt and E. Eibl-Eibesfeldt, "Das Parasitenabwehren der Minima-Arbeiterinnen der Blatt-schneider-Ameise (*Atta cephalotes*)," *Zeitschrift für Tierpsychologie* 24(3): 278–281 (1967); D. H. Feener and K. A. G. Moss, "Defense against parasites by hitchhikers in leaf-cutting ants: a quantitative assessment," *Behavioral Ecology and Sociobiology* 26(1): 17–26 (1990); T. A. Linksvayer, A. C. McCall, R. M. Jensen, C. M. Marshall, J. W. Miner, and M. J. McKone, "The function of hitchhiking behavior in the leaf-cutting ant *Atta cephalotes*," *Biotropica* 34(1): 93–100 (2002); E. H. M. Vieira-Neto, F. M. Mundim, and H. L. Vasconcelos, "Hitchhiking behaviour in leaf-cutter ants: an experimental evaluation of three hypotheses," *Insectes Sociaux* 53(3): 326–332 (2006).

52 | M. E. A. Whitehouse and K. Jaffe, "Ant wars: combat strategies, territory and nest defence in the leaf-cutting ant *Atta laevigata*," *Animal Behaviour* 51(6): 1207–1217 (1996).

53 | W. O. H. Hughes and D. Goulson, "Polyethism and the importance of context in the alarm reaction of the grass-cutting ant, *Atta capiguara*," *Behavioral Ecology and Sociobiology* 49(6): 503–508 (2001).

54 | S. Powell and E. Clark, "Combat between large derived societies: a subterranean army ant established as a predator of mature leaf-cutting ant colonies," *Insectes Sociaux* 51(4): 342–351 (2004).

Chapter 6: Harvesting Vegetation

55 | C. M. Nichols-Orians and J. C. Schultz, "Leaf toughness affects leaf harvesting by the leaf-cutter ant, *Atta cephalotes* (L.) (Hymenoptera: Formicidae)," *Biotropica* 21(1): 80–83 (1989). For a review, see R. Wirth, H. Herz, R. J. Ryel, W. Beyschlag, and B. Hölldobler, *Herbivory of Leaf-Cutting Ants: A Case Study on* Atta colombica *in the Tropical Rainforest of Panama* (New York: Springer-Verlag, 2003).

56 | J. M. van Breda and D. J. Stradling, "Mechanisms affecting load size determination in *Atta cephalotes* (L.) (Hymenoptera, Formicidae)," *Insectes Sociaux* 41(4): 423–434 (1994).

57 | H. Helanterä, and F. L. W. Ratnieks, "Geometry explains the benefits of division of labour in a leafcutter ant," *Proceedings of the Royal Society B* 275: 1255–1260 (2008).

58 | H. Markl, "Stridulation in leaf-cutting ants," *Science* 149: 1392–1393 (1965); H. Markl, "Die Verständigung durch Stridulationssignale bei Blattschneiderameisen, II: Erzeugung und Eigenschaften der Signale," *Zeitschrift für vergleichende Physiologie* 60(2): 103–150 (1968).

59 | J. Tautz, F. Roces, and B. Hölldobler, "Use of a sound-based vibratome by leaf-cutting ants," *Science* 267: 84–87 (1995).

60 | F. Roces and J. R. B. Lighton, "Larger bites of leaf-cutting ants," *Nature* 373: 392–393 (1995).

61 | J. K. Wetterer, "Forager polymorphism, size-matching, and load delivery in the leaf-cutting ant, *Atta cephalotes*," *Ecological Entomology* 19(1): 57–64 (1994); D. A. Waller,

"The foraging ecology of *Atta texana* in Texas," in C. S. Lofgren and R. K. Vander Meer, eds., *Fire Ants and Leaf-Cutting Ants: Biology and Management* (Boulder, CO: Westview Press, 1986), pp. 146–158.

62 | See R. Wirth, H. Herz, R. J. Ryel, W. Beyschlag, and B. Hölldobler, *Herbivory of Leaf-Cutting Ants: A Case Study on* Atta colombica *in the Tropical Rainforest of Panama* (New York: Springer-Verlag, 2003).

63 | M. Burd, "Variable load size-ant size matching in leaf-cutting ants, *Atta colombica* (Hymenoptera: Formicidae)," *Journal of Insect Behavior* 8(5): 715–722 (1995); M. Burd, "Foraging performance by *Atta colombica,* a leaf-cutting ant," *American Naturalist* 148(4): 597–612 (1996).

64 | F. Roces and B. Hölldobler, "Leaf density and a trade-off between load size selection and recruitment behavior in the ant *Atta cephalotes*," *Oecologia* 97(1): 1–8 (1994); F. Roces and J. A. Núñez, "Information about food quality influences load-size selection in recruited leaf-cutting ants," *Animal Behaviour* 45(1): 135–143 (1993).

65 | For a detailed discussion of these issues, see M. Burd, "Server system and queuing models of leaf harvesting by leaf-cutting ants," *American Naturalist* 148(4): 613–629 (1996). For a recent review see F. Roces and M. Bollazzi, "Information transfer and the organization of foraging in grass- and leaf-cutting ants," in S. Jarau and M. Hrncir, eds., *Food Exploitation by Social Insects* (Boca Raton: CRC Press, 2009), pp. 261–275.

66 | S. T. Meyer, F. Roces, and R. Wirth, "Selecting the drought stressed: effects of plant stress on intraspecific and within-plant herbivory patterns of the leaf-cutting ant *Atta colombica*," *Functional Ecology* 20(6): 973–981 (2006).

67 | B. Hölldobler and E. O. Wilson, *The Ants* (Cambridge, MA: Belknap Press of Harvard University Press, 1990).

68 | C. Anderson and J. L. V. Jadin, "The adaptive benefit of leaf transfer in *Atta colombica*," *Insectes Sociaux* 48(4): 404–405 (2001); A. G. Hart and F. L. W. Ratnieks, "Leaf caching in the leafcutting ant *Atta colombica*: organizational shift, task partitioning and making the best of a bad job," *Animal Behaviour* 62(2): 227–234 (2001).

69 | S. P. Hubbell, L. K. Johnson, E. Stanislav, B. Wilson, and H. Fowler, "Foraging by bucket-brigade in leaf-cutter ants," *Biotropica* 12(3): 210–213 (1980).

70 | J. Röschard and F. Roces, "The effect of load length, width and mass on transport rate in the grass-cutting ant *Atta vollenweideri*," *Oecologia* 131(2): 319–324 (2002); J. Röschard and F. Roces, "Cutters, carriers and transport chains: distance-dependent foraging strategies in the grass-cutting ant *Atta vollenweideri*," *Insectes Sociaux* 50(3): 237–244 (2003).

71 | F. Roces and J. R. B. Lighton, "Larger bites of leaf-cutting ants," *Nature* 373: 392–393 (1995).

72 | S. P. Hubbell, L. K. Johnson, E. Stanislav, B. Wilson, and H. Fowler, "Foraging by bucket-brigade in leaf-cutter ants," *Biotropica* 12(3): 210–213 (1980); H. G. Fowler and S. W. Robinson, "Foraging by *Atta sexdens* (Formicidae: Attini): seasonal patterns, caste and efficiency," *Ecological Entomology* 4(3): 239–247 (1979); C. Anderson and F. L. W.

Ratnieks, "Task partitioning in insect societies, I: Effect of colony size on queuing delay and colony ergonomic efficiency," *American Naturalist* 154(5): 521–535 (1999); A. G. Hart and F. L. W. Ratnieks, "Leaf caching in the *Atta* leaf-cutting ant *Atta colombica:* organizational shift, task partitioning and making the best of a bad job," *Animal Behaviour* 62(2): 227–234 (2001).

73 | C. Anderson, J. J. Boomsma, and J. J. Bartholdi III, "Task partitioning in insect societies: bucket brigades," *Insectes Sociaux* 49(2): 171–180 (2002).

74 | J. Röschard and F. Roces, "The effect of load length, width and mass on transport rate in the grass-cutting ant *Atta vollenweideri*," *Oecologia* 131(2): 319–324 (2002); J. Röschard and F. Roces, "Cutters, carriers and transport chains: distance-dependent foraging strategies in the grass-cutting ant *Atta vollenweideri*," *Insectes Sociaux* 50(3): 237–244 (2003).

75 | F. Roces, "Individual complexity and self-organization in foraging by leaf-cutting ants," *Biological Bulletin* 202(3): 306–313 (2002).

76 | J. Röschard and F. Roces, "The effect of load length, width and mass on transport rate in the grass-cutting ant *Atta vollenweideri*," *Oecologia* 131(2): 319–324 (2002); J. Röschard and F. Roces, "Cutters, carriers and transport chains: distance-dependent foraging strategies in the grass-cutting ant *Atta vollenweideri*," *Insectes Sociaux* 50(3): 237–244 (2003).

77 | C. Anderson, J. J. Boomsma, and J. J. Bartholdi III, "Task partitioning in insect societies: bucket brigades," *Insectes Sociaux* 49(2): 171–180 (2002).

78 | J. J. Howard, "Leaf-cutting and diet selection: relative influence of leaf chemistry and physical factors," *Ecology* 69(1): 250–260 (1988).

79 | C. M. Nichols-Orians and J. C. Schultz, "Interactions among leaf toughness, chemistry, and harvesting by attine ants," *Ecological Entomology* 15(3): 311–320 (1990).

80 | There exists a rich and sometimes contradicting literature on food plant selection in attine ants, which is partially reviewed in R. Wirth, H. Herz, R. J. Ryel, W. Beyschlag, and B. Hölldobler, *Herbivory of Leaf-Cutting Ants: A Case Study on* Atta colombica *in the Tropical Rainforest of Panama* (New York: Springer-Verlag, 2003).

Chapter 7: Communication in *Atta*

81 | J. C. Moser and M. S. Blum, "Trail marking substance of the Texas leaf-cutting ant: source and potency," *Science* 140: 1228 (1963).

82 | K. Jaffe and P. E. Howse, "The mass recruitment system of the leaf-cutting ant *Atta cephalotes* (L.)," *Animal Behaviour* 27(3): 930–939 (1979); B. Hölldobler and E. O. Wilson, *The Ants* (Cambridge, MA: Belknap Press of Harvard University Press, 1990).

83 | J. H. Tumlinson, R. M. Silverstein, J. C. Moser, R. G. Brownlee, and J. M. Ruth, "Identification of the trail pheromone of a leaf-cutting ant, *Atta texana*," *Nature* 234: 348–349 (1971).

84 | J. H. Cross, R. C. Byler, U. Ravid, R. M. Silverstein, S. W. Robinson, P. M. Baker, J. S. De Oliveira, A. R. Jutsum, and J. M. Cherrett, "The major component of the trail

pheromone of the leaf-cutting ant, *Atta sexdens rubropilosa* Forel: 3-ethyl-2,5-dimethylpyr-azine," *Journal of Chemical Ecology* 5: 187–203 (1979).

85 | R. G. Riley, R. M. Silverstein, B. Carroll, and R. Carroll, "Methyl 4-methylpyrrole-2-carboxylate: a volatile trail pheromone from the leaf-cutting ant, *Atta cephalotes*," *Journal of Insect Physiology* 20(4): 651–654 (1974).

86 | C. J. Kleineidam, W. Rössler, B. Hölldobler, and F. Roces, "Perceptual differences in trail-following leaf-cutting ants relate to body size," *Journal of Insect Physiology* 53(12): 1233–1241 (2007).

87 | C. Kelber, W. Rössler, F. Roces, and C. J. Kleineidam, "The antennal lobes of fungus-growing ants (Attini): neuroanatomical traits and evolutionary trends," *Brain, Behavior, and Evolution* 73: 273–284 (2009).

88 | F. Roces and B. Hölldobler, "Leaf density and a trade-off between load-size selection and recruitment behavior in the ant *Atta cephalotes*," *Oecologia* 97(1): 1–8 (1994); C. M. Nichols-Orians and J. C. Schultz, "Interactions among leaf toughness, chemistry, and har-vesting by attine ants," *Ecological Entomology* 15(3): 311–320 (1990).

89 | J. W. S. Bradshaw, P. E. Howse, and R. Baker, "A novel autostimulatory phero-mone regulating transport of leaves in *Atta cephalotes*," *Animal Behaviour* 34(1): 234–240 (1986); B. Hölldobler and E. O. Wilson, "Nest area exploration and recognition in leaf-cutter ants (*Atta cephalotes*)," *Journal of Insect Physiology* 32(2): 143–150 (1986).

90 | F. Roces, J. Tautz, and B. Hölldobler, "Stridulation in leaf-cutting ants: short-range recruitment through plant-borne vibrations," *Naturwissenschaften* 80(11): 521–524 (1993).

91 | H. Markl, "Stridulation in leaf-cutting ants," *Science* 149: 1392–1393 (1965); H. Markl, "Die Verständigung durch Stridulationssignale bei Blattschneiderameisen, II: Erzeugung und Eigenschaften der Signale," *Zeitschrift für vergleichende Physiologie* 60(2): 103–150 (1968).

92 | F. Roces, J. Tautz, and B. Hölldobler, "Stridulation in leaf-cutting ants: short-range recruitment through plant-borne vibrations," *Naturwissenschaften* 80(11): 521–524 (1993); B. Hölldobler and F. Roces, "The behavioral ecology of stridulatory communi-cation in leaf-cutting ants," in L. A. Dugkatin, ed., *Model Systems in Behavioral Ecology: Integrating Conceptual, Theoretical, and Empirical Approaches* (Princeton, NJ: Princeton University Press, 2001), pp. 92–109.

93 | F. Roces and B. Hölldobler, "Use of stridulation in foraging leaf-cutting ants: mechanical support during cutting or short-range recruitment signal?" *Behavioral Ecology and Sociobiology* 39(5): 293–299 (1996).

94 | F. Roces, personal communication.

95 | F. Roces and B. Hölldobler, "Vibrational communication between hitchhikers and foragers in leaf-cutting ants (*Atta cephalotes*)," *Behavioral Ecology and Sociobiology* 37(5): 297–302 (1995).

96 | H. Markl, "Stridulation in leaf-cutting ants," *Science* 149: 1392–1393 (1965); and H. Markl, "Die Verständigung durch Stridulationssignale bei Blattschneiderameisen, II:

Erzeugung und Eigenschaften der Signale," *Zeitschrift für vergleichende Physiologie* 60(2): 103–150 (1968).

97 | A. Butenandt, B. Linzen, and M. Lindauer, "Über einen Duftstoff aus der Mandibeldrüse der Blatt-schneiderameise *Atta sexdens rubropilosa* Forel," *Archives d'Anatomie Microscopique et de Morphologie Expérimentale* 48(Supplement): 13–19 (1959).

98 | M. S. Blum, F. Padovani, and E. Amante, "Alkanones and terpenes in the mandibular glands of *Atta* species (Hymenoptera: Formicidae)," *Comparative Biochemistry and Physiology* 26: 291–299 (1968).

99 | R. R. Do Nascimento, E. D. Morgan, J. Billen, E. Schoeters, T. M. C. Della Lucia, and J. M. S. Bento, "Variation with caste of the mandibular gland secretion in the leaf-cutting ant *Atta sexdens rubropilosa*," *Journal of Chemical Ecology* 19(5): 907–918 (1993).

100 | For additional information on caste variation of pheromones in *Atta*, see W. O. H. Hughes, P. E. Howse, and D. Goulson, "Mandibular gland chemistry of grass-cutting ants: species, caste, and colony variation," *Journal of Chemical Ecology* 27(1): 109–124 (2001).

101 | A. Endler, J. Liebig, T. Schmitt, J. E. Parker, G. R. Jones, P. Schreier, and B. Hölldobler, "Surface hydrocarbons of queen eggs regulate worker reproduction in a social insect," *Proceedings of the National Academy of Sciences USA* 101(9): 2945–2950 (2004); A. Endler, J. Liebig, and B. Hölldobler, "Queen fertility, egg marking and colony size in the ant *Camponotus floridanus*," *Behavioral Ecology and Sociobiology* 59(4): 490–499 (2006).

Chapter 8: The Ant-Fungus Mutualism

102 | M. Littledyke and J. M. Cherrett, "Defence mechanisms in young and old leaves against cutting by the leaf-cutting ants *Atta cephalotes* (L.) and *Acromyrmex octospinosus* (Reich) (Hymenoptera: Formicidae)," *Bulletin of Entomological Research* 68(2): 263–271 (1978); S. P. Hubell, D. F. Wiemer, and A. Adejare, "An antifungal terpenoid defends a Neotropical tree (Hymenaea) against attack by fungus-growing ants (*Atta*)," *Oecologia* 60(3): 321–327 (1983); J. J. Howard, "Leafcutting and diet selection: relative influence of leaf-chemistry and physical features," *Ecology* 69(1): 250–260 (1988). See a review in R. Wirth, H. Herz, R. J. Ryel, W. Beyschlag, and B. Hölldobler, *Herbivory of Leaf-Cutting Ants: A Case Study on* Atta colombica *in the Tropical Rainforest of Panama* (New York: Springer-Verlag, 2003).

103 | J. J. Knapp, P. E. Howse, and A. Kermarrec, "Factors controlling foraging patterns in the leaf-cutting ant *Acromyrmex octospinosus* (Reich)," in R. K. Vander Meer, K. Jaffe, and A. Cedeno, eds., *Applied Myrmecology: A World Perspective* (Boulder, CO: Westview Press, 1990), pp. 382–409; H. L. Vasconcelos and H. G. Fowler, "Foraging and fungal substrate selection by leaf-cutting ants," in R. K. Vander Meer, K. Jaffe, and A. Cedeno, eds., *Applied Myrmecology: A World Perspective* (Boulder, CO: Westview Press, 1990), pp. 410–419.

104 | P. Ridley, P. E. Howse, and C. W. Jackson, "Control of the behaviour of leaf-cutting ants by their 'symbiotic' fungus," *Experientia* 52(6): 631–635 (1996).

105 | R. D. North, C. W. Jackson, and P. E. Howse, "Communication between the fungus garden and workers of the leaf-cutting ant, *Atta sexdens rubropilosa,* regarding choice of substrate for the fungus," *Physiological Entomology* 24(2): 127–133 (1999).

106 | R. D. North, C. W. Jackson, and P. E. Howse, "Communication between the fungus garden and workers of the leaf-cutting ant, *Atta sexdens rubropilosa,* regarding choice of substrate for the fungus," *Physiological Entomology* 24(2): 127–133 (1999).

107 | F. Roces, "Olfactory conditioning during the recruitment process in a leaf-cutting ant," *Oecologia* 83(2): 261–262 (1990); F. Roces, "Odour learning and decision-making during food collection in the leaf-cutting ant *Acromyrmex lundi,*" *Insectes Sociaux* 41(3): 235–239 (1994); J. J. Howard, L. Henneman, G. Cronin, J. A. Fox, and G. Hormiga, "Conditioning of scouts and recruits during foraging by a leaf-cutting ant, *Atta colombica,*" *Animal Behaviour* 52(2): 299–306 (1996).

108 | F. Roces, "Odour learning and decision-making during food collection in the leaf-cutting ant *Acromyrmex lundi,*" *Insectes Sociaux* 41(3): 235–239 (1994).

109 | J. J. Howard, M. L. Henneman, G. Cronin, J. A. Fox, and G. Hormiga, "Conditioning of scouts and recruits during foraging by a leaf-cutting ant, *Atta colombica,*" *Animal Behaviour* 52(2): 299–306 (1996).

110 | H. Herz, B. Hölldobler, and F. Roces, "Delayed rejection in a leaf-cutting ant after foraging on plants unsuitable for the symbiotic fungus," *Behavioral Ecology* 19(3): 575–582 (2008).

111 | A. N. M. Bot, S. A. Rehner, and J. J. Boomsma, "Partial incompatibility between ants and symbiotic fungi in two sympatric species of *Acromyrmex* leaf-cutting ants," *Evolution* 55(10): 1980–1991 (2001).

112 | M. Poulsen and J. J. Boomsma, "Mutualistic fungi control crop diversity in fungus-growing ants," *Science* 307: 741–744 (2005).

113 | M. Poulsen and J. J. Boomsma, "Mutualistic fungi control crop diversity in fungus-growing ants," *Science* 307: 741–744 (2005).

114 | For a more detailed review and excellent discussion of issues concerning ant-fungus conflict, see also U. G. Mueller, "Ant versus fungus versus mutualism: ant-cultivar conflict and the deconstruction of the attine ant-fungus symbiosis," *American Naturalist* 160(Supplement): S67–S98 (2002).

115 | J. N. Seal and W. R. Tschinkel, "Co-evolution and the superorganism: Switching cultivars does not alter the performance of fungus-gardening ant colonies," *Functional Ecology* 21(5): 988–997 (2007).

Chapter 9: Hygiene in the Symbiosis

116 | For a review, see B. Hölldobler and E. O. Wilson, *The Ants* (Cambridge, MA: Belknap Press of Harvard University Press, 1990); and R. Wirth, H. Herz, R. J. Ryel, W. Beyschlag, and B. Hölldobler, *Herbivory of Leaf-Cutting Ants: A Case Study on* Atta colombica *in the Tropical Rainforest of Panama* (New York: Springer-Verlag, 2003).

117 | U. Maschwitz, K. Koob, and H. Schildknecht, "Ein Beitrag zur Funktion der Metathorakaldrüse der Ameisen," *Journal of Insect Physiology* 16(2): 387–404 (1970); U. Maschwitz, "Vergleichende Untersuchungen zur Funktion der Ameisenmetathorakaldrüse," *Oecologia* 16(4): 303–310 (1974).

118 | H. Schildknecht and K. Koob, "Plant bioregulators in the metathoracic glands of myrmicine ants," *Angewandte Chemie* 9(2): 173 (1970); H. Schildknecht and K. Koob, "Myrmicacin, the first insect herbicide," *Angewandte Chemie* 10(2): 124–125 (1971).

119 | D. Ortius-Lechner, R. Maile, E. D. Morgan, and J. J. Boomsma, "Metapleural gland secretion of the leaf-cutter ant *Acromyrmex octospinosus:* new compounds and their functional significance," *Journal of Chemical Ecology* 26(7): 1667–1683 (2000).

120 | E. O. Wilson, "Caste and division of labor in leaf-cutter ants (Hymenoptera: Formicidae: *Atta*), I: The overall pattern in *A. sexdens,*" *Behavioral Ecology and Sociobiology* 7(2): 143–156 (1980); E. O. Wilson, "Caste and division of labor in leaf-cutter ants (Hymenoptera: Formicidae: *Atta*), II: The ergonomic optimization of leaf cutting," *Behavioral Ecology and Sociobiology* 7(2): 157–165 (1980); E. O. Wilson, "Caste and division of labor in leaf-cutter ants (Hymenoptera: Formicidae: *Atta*), III: Ergonomic resiliency in foraging by *A. cephalotes,*" *Behavioral Ecology and Sociobiology* 14(1): 47–54 (1983); E. O. Wilson, "Caste and division of labor in leaf-cutter ants (Hymenoptera: Formicidae: *Atta*), IV: Colony ontogeny of *A. cephalotes,*" *Behavioral Ecology and Sociobiology* 14(1): 55–60 (1983); A. N. M. Bot, M. L. Obermayer, B. Hölldobler, and J. J. Boomsma, "Functional morphology of the metapleural gland in the leaf-cutting ant *Acromyrmex octospinosus,*" *Insectes Sociaux* 48(1): 63–66 (2001).

121 | C. R. Currie, "Prevalence and impact of a virulent parasite on a tripartite mutualism," *Oecologia* 128: 99–106 (2001). For an excellent review, see C. R. Currie, "A community of ants, fungi, and bacteria: a multilateral approach to studying symbiosis," *Annual Review of Microbiology* 55: 357–380 (2001).

122 | R. J. Powell and D. J. Stradling, "Factors influencing the growth of *Attamyces bromatificus*, a symbiont of attine ants," *Transactions of the British Mycological Society* 87(2): 205–213 (1986).

123 | D. Ortius-Lechner, R. Maile, E. D. Morgan, and J. J. Boomsma, "Metapleural gland secretion of the leaf-cutter ant, *Acromyrmex octospinosus:* new compounds and their functional significance," *Journal of Chemical Ecology* 26(7): 1667–1683 (2000).

124 | C. R. Currie, U. G. Mueller, and D. Malloch, "The agricultural pathology of ant fungus gardens," *Proceedings of the National Academy of Sciences USA* 96(14): 7998–8002 (1999).

125 | A. S. Mikheyev, U. G. Mueller, and P. Abbot, "Cryptic sex and many-to-one coevolution in the fungus-growing ant symbiosis," *Proceedings of the National Academy of Sciences USA* 103(28): 10702–10706 (2006).

126 | Originally, this actinomycete was thought to be of the genus *Streptomyces* (Streptomycetaceae: Actinomycetes). This identification appears to be incorrect (R. Wirth, personal communication), and ongoing molecular phylogenetic analyses have revealed that the symbiotic bacterium belongs to the actinomycetous family Pseudonocardiaceae (C. R. Currie, personal communication). See also corrigendum in *Nature* 423: 461 (2003).

127 | C. R. Currie, J. A. Scott, R. C. Summerbell, and D. Malloch, "Fungus-growing ants use antibiotic-producing bacteria to control garden parasites," *Nature* 398: 701–704 (1999).

128 | D.-C. Oh, M. Poulsen, C. R. Currie, and J. Clardy, "Dentigerumycin: a bacterial mediator of ant-fungus symbiosis," *Nature Chemical Biology* 5: 391–393 (2009).

129 | C. R. Currie, M. Poulsen, J. Mendenhall, J. J. Boomsma, and J. Billen, "Coevolved crypts and exocrine glands support mutualistic bacteria in fungus-growing ants," *Science* 311: 81–83 (2006).

130 | A. E. F. Little, T. Murakami, U. G. Mueller, and C. R. Currie, "The infrabuccal pellet piles of fungus-growing ants," *Naturwissenschaften* 90: 558–562 (2003); A. E. F. Little, T. Murakami, U. G. Mueller, and C. R. Currie, "Defending against parasites: fungus-growing ants combine specialized behaviours and microbial symbionts to protect their fungus gardens," *Biology Letters* 2(1): 12–16 (2006).

131 | R. Wirth, H. Herz, R. J. Ryel, W. Beyschlag, and B. Hölldobler, *Herbivory of Leaf-Cutting Ants: A Case Study on* Atta colombica *in the Tropical Rainforest of Panama* (New York: Springer-Verlag, 2003).

132 | C. Kost, T. Lakatos, I. Bötcher, W.-R. Arendholz, M. Redenbach, and R. Wirth, "Nonspecific association between filamentous bacteria and fungus-growing ants," *Naturwissenschaften* 94: 821–828 (2007).

133 | S. Haeder, R. Wirth, H. Herz, and D. Spiteller, "Candicidin-producing *Streptomyces* support leaf-cutting ants to protect their fungus garden against the pathogenic fungus *Escovopsis*," *Proceedings of the National Academy of Sciences USA* 106(12): 4742–4746 (2008).

134 | E. J. Caldera, M. Poulsen, G. Suen, and C. R. Currie, "Insect symbiosis: a case study of past, present, and future fungus-growing ant research," *Environmental Entomology* 38: 78–82 (2009).

135 | C. R. Currie, U. G. Mueller, and D. Malloch, "The agricultural pathology of ant fungus gardens," *Proceedings of the National Academy of Sciences USA* 96(14): 7998–8002 (1999).

136 | A. E. F. Little and C. R. Currie, "Black yeast symbionts compromise the efficiency of antibiotic defenses in fungus-growing ants," *Ecology* 89: 1216–1222 (2008).

137 | A. A. Pinto-Tomás, M. A. Anderson, G. Suen, D. M. Stevenson, F. A. T. Chu, W. W. Cleland, P. J. Weimer, and C. R. Currie, "Symbiotic nitrogen fixation in the fungus gardens of leaf-cutter ants," *Science* 326: 1120–1123 (2009).

Chapter 10: Waste Management

138 | J. A. Zeh, A. D. Zeh, and D. W. Zeh, "Dump material as an effective small-scale deterrent to herbivory by *Atta cephalotes*," *Biotropica* 31(2): 368–371 (1999); C. R. Currie, J. A. Scott, R. C. Summerbell, and D. Malloch, "Fungus-growing ants use antibiotic-producing bacteria to control garden parasites," *Nature* 398: 701–704 (1999).

139 | C. Anderson and F. L. W. Ratnieks, "Task partitioning in insect societies: novel situations," *Insectes Sociaux* 47(2): 198–199 (2000).

140 | C. Anderson and F. L. W. Ratnieks, "Task partitioning in insect societies: novel situations," *Insectes Sociaux* 47(2): 198–199 (2000).

141 | A. N. M. Bot, C. R. Currie, A. G. Hart, and J. J. Boomsma, "Waste management in leaf-cutting ants," *Ethology Ecology & Evolution* 13(3): 225–237 (2001); A. G. Hart and F. L. W. Ratnieks, "Waste management in the leaf-cutting ant *Atta colombica*," *Behavioral Ecology* 13(2): 224–231 (2002).

142 | B. Hölldobler and E. O. Wilson, *The Ants* (Cambridge, MA: Belknap Press of Harvard University Press, 1990).

143 | R. M. M. Adams, U. G. Mueller, A. K. Holloway, A. M. Green, and J. Narozniak, "Garden sharing and garden stealing in fungus-growing ants," *Naturwissenschaften* 87(11): 491–493 (2000).

144 | M. Autuori, "Contribuição para o conhecimento da saúva *(Atta* spp.—Hymenoptera—Formicidae), V: Número de formas aladas e redução dos sauveiros iniciais," *Arquivos do Instituto Biológico São Paulo* 19(22): 325–331 (1950).

145 | S. W. Rissing, G. B. Pollock, M. R. Higgins, R. H. Hagen, and D. R. Smith, "Foraging specialization without relatedness or dominance among co-founding ant queens," *Nature* 338: 420–422 (1989).

146 | In this context the new findings on long-distance horizontal transmission of fungal symbionts between leafcutter ants are of particular interest. See A. S. Mikheyev, U. G. Mueller, and P. Abbot, "Cryptic sex and many-to-one co-evolution in the fungus-growing ant symbiosis," *Proceedings of the National Academy of Sciences USA* 103(28): 10702–10706 (2006).

Chapter 11: Agropredators and Agroparasites

147 | M. B. Dijkstra and J. J. Boomsma, "*Gnamptogenys hartmani* Wheeler (Ponerinae: Ectatommini): an agro-predator of *Trachymyrmex* and *Sericomyrmex* fungus-growing ants," *Naturwissenschaften* 90(12): 568–571 (2003).

148 | R. M. M. Adams, U. G. Mueller, A. K. Holloway, A. M. Green, and J. Narozniak, "Garden sharing and garden stealing in fungus-growing ants," *Naturwissenschaften* 87(11): 491–493 (2000); R. M. M. Adams, U. G. Mueller, T. R. Schultz, and B. Norden, "Agro-predation: usurpation of attine fungus gardens by *Megalomyrmex* ants," *Naturwissenschaften* 87(12): 549–554 (2000); W. M. Wheeler, "A new guest-ant and other new Formicidae from Barro Colorado Island, Panama," *Biological Bulletin* 49(1): 150–181 (1925).

149 | C. R. F. Brandão, "Systematic revision of the Neotropical ant genus *Megalomyrmex* Forel (Hymenoptera, Formicidae, Myrmicinae), with the description of thirteen new species," *Arquivos de Zoologia* (São Paulo) 31: 411–481 (1990).

150 | T. R. Schultz, D. Bekkevold, and J. J. Boomsma, "*Acromyrmex insinuator* new species: an incipient social parasite of fungus-growing ants," *Insectes Sociaux* 45(4): 457–471 (1998).

Chapter 12: Leafcutter Nests

151 | See B. Hölldobler and E. O. Wilson, *The Ants* (Cambridge, MA: Belknap Press of Harvard University Press, 1990); colonies of *Acromyrmex* species are correspondingly smaller and less complex.

152 | A. A. Moreira, L. C. Forti, A. P. P. Andrade, M. A. C. Boaretto, and J. F. S. Lopes, "Nest architecture of *Atta laevigata* (F. Smith, 1858) (Hymenoptera: Formicidae)," *Studies on Neotropical Fauna and Environment* 39(2): 109–116 (2004); A. A. Moreira, L. C. Forti, M. A. C. Boaretto, A. P. P. Andrade, J. F. S. Lopes, and V. M. Ramos, "External and internal structure of *Atta bisphaerica* Forel (Hymenoptera: Formicidae) nests," *Journal of Applied Entomology* 128(3): 204–211 (2004).

153 | L. C. Forti and F. Roces, personal communication.

154 | For a discussion, see A. A. Moreira, L. C. Forti, A. P. P. Andrade, M. A. C. Boaretto, and J. F. S. Lopes, "Nest architecture of *Atta laevigata* (F. Smith, 1858) (Hymenoptera: Formicidae)," *Studies on Neotropical Fauna and Environment* 39(2): 109–116 (2004); and A. A. Moreira, L. C. Forti, M. A. C. Boaretto, A. P. P. Andrade, J. F. S. Lopes, and V. M. Ramos, "External and internal structure of *Atta bisphaerica* Forel (Hymenoptera: Formicidae) nests," *Journal of Applied Entomology* 128(3): 204–211 (2004).

155 | C. Kleineidam and J. Tautz, "Perception of carbon dioxide and other 'air-condition' parameters in the leaf-cutting ant *Atta cephalotes*," *Naturwissenschaften* 83(12): 566–568 (1996).

156 | C. Kleineidam and F. Roces, "Carbon dioxide concentrations and nest ventilation in nests of the leaf-cutting ant *Atta vollenweideri*," *Insectes Sociaux* 47(3): 241–248 (2000); C. Kleineidam, R. Ernst, and F. Roces, "Wind-induced ventilation of the giant nests of the leaf-cutting ant *Atta vollenweideri*," *Naturwissenschaften* 88(7): 301–305 (2001).

157 | F. Roces, personal communication.

158 | F. Roces and C. Kleineidam, "Humidity preference for fungus culturing by workers of the leaf-cutting ant *Atta sexdens rubropilosa*," *Insectes Sociaux* 47(4): 348–350 (2000).

159 | M. Bollazzi and F. Roces, "To build or not to build: circulating dry air organizes collective building for climate control in the leaf-cutting ant *Acromyrmex ambiguus*," *Animal Behaviour* 74(5): 1349–1355 (2007).

160 | M. Bollazzi, J. Kronenbitter, and F. Roces, "Soil temperature, digging behaviour, and adaptive value of nest depth in South American species of *Acromyrmex* leaf-cutting ants," *Oecologia* 158: 165–175 (2008).

161 | C. J. Kleineidam, M. Ruchty, Z. A. Casero-Montes, and F. Roces, "Thermal radiation as a learned orientation cue in leaf-cutting ants (*Atta vollenweideri*)," *Journal of Insect*

Physiology 53: 478–487 (2007); M. Ruchty, R. Romani, L. S. Kuebler, S. Russchioni, F. Roces, N. Isidoro, and C. Kleineidam, "The thermo-sensitive-*sensilla coeloconica* of leaf-cutting ants (*Atta vollenweideri*)," *Arthropod Structure of Development* 38: 195–205 (2009).

Chapter 13: Trail and Trunk Routes

162 | A. A. Moreira, L. C. Forti, A. P. P. Andrade, M. A. C. Boaretto, and J. F. S. Lopes, "Nest architecture of *Atta laevigata* (F. Smith, 1858) (Hymenoptera: Formicidae)," *Studies on Neotropical Fauna and Environment* 39(2): 109–116 (2004).

163 | F. Roces, personal communication.

164 | L. L. Rockwood and S. P. Hubbell, "Host-plant selection, diet diversity, and optimal foraging in a tropical leafcutting ant," *Oecologia* 74(1): 55–61 (1987).

165 | H. G. Fowler and E. W. Stiles, "Conservative resource management by leaf-cutting ants? The role of foraging territories and trails, and environmental patchiness," *Sociobiology* 5(1): 25–41 (1980).

166 | J. J. Howard, "Costs of trail construction and maintenance in the leaf-cutting ant *Atta colombica*," *Behavioral Ecology and Sociobiology* 49(5): 348–356 (2001).

167 | R. Wirth, H. Herz, R. J. Ryel, W. Beyschlag, and B. Hölldobler, *Herbivory of Leaf-Cutting Ants: A Case Study on* Atta colombica *in the Tropical Rainforest of Panama* (New York: Springer-Verlag, 2003). See also C. Kost, E. G. de Oliveira Kost, T. A. Knoch, and R. Wirth, "Spatio-temporal permanence and plasticity of foraging trails in young and mature leaf-cutting ant colonies (*Atta* spp.)," *Journal of Tropical Ecology* 21(6): 677–688 (2005).

168 | A. A. Moreira, L. C. Forti, A. P. P. Andrade, M. A. C. Boaretto, and J. F. S. Lopes, "Nest architecture of *Atta laevigata* (F. Smith, 1858) (Hymenoptera: Formicidae)," *Studies on Neotropical Fauna and Environment* 39(2): 109–116 (2004); A. A. Moreira, L. C. Forti, M. A. C. Boaretto, A. P. P. Andrade, J. F. S. Lopes, and V. M. Ramos, "External and internal structure of *Atta bisphaerica* Forel (Hymenoptera: Formicidae) nests," *Journal of Applied Entomology* 128(3): 204–211 (2004); L. C. Forti and F. Roces, personal communication; N. A. Weber, *Gardening Ants: The Attines* (Philadelphia: American Philosophical Society, 1972); J. C. Moser, "Contents and structure of *Atta texana* nest in summer," *Annals of the Entomological Society of America* 56(3): 286–291 (1963).

169 | T. D. Seeley, *The Wisdom of the Hive: The Social Physiology of Honey Bee Colonies* (Cambridge, MA: Harvard University Press, 1995).

170 | D. S. Wilson and E. Sober, "Reviving the superorganism," *Journal of Theoretical Biology* 136(3): 337–356 (1989).

171 | B. Hölldobler and E. O. Wilson, *The Ants* (Cambridge, MA: Belknap Press of Harvard University Press, 1990).

INDEX

ABOUT THE AUTHORS

Bert Hölldobler is Foundation Professor of Life Sciences at Arizona State University. Before joining ASU he was the Alexander Agassiz Professor of Zoology at Harvard University (1973–1990), and he held the Chair of Behavioral Physiology and Sociobiology at the University of Würzburg, in Germany (1989–2004). In 2002 he was appointed Andrew D. White Professor at Large at Cornell University.

He is a member of several national and international academies, among them the German Academy of Sciences (Leopoldina), the American Philosophical Society, the American Academy of Arts and Sciences, and the National Academy of Sciences (USA). He is the author of several books, including *The Ants,* which he coauthored with Edward O. Wilson, and for which they received a Pulitzer Prize (1991) for nonfiction writing, and their book *Journey to the Ants,* which was awarded the Phi Beta Kappa Prize. He is the recipient of some of the most prestigious research prizes in Germany, among others the Gottfried Wilhelm Leibniz Prize of the German Science Foundation, the Körber Prize for the European Sciences, and the Alfried Krupp Science Prize. At ASU Bert Hölldobler is a cofounder of the new Center for Social Dynamics and Complexity, and he plays a key role in organizing the new social insect research group at the School of Life Sciences. Along with his wife, he divides his time between Arizona and Germany.

Edward O. Wilson was born in Birmingham, Alabama, in 1929 and was drawn to the natural environment from a young age. After studying evolutionary biology at the University of Alabama, he has spent his career focused on scientific research and teaching, including forty-one years on the faculty of Harvard University. His twenty books and more than four hundred mostly technical articles have won him over one hundred awards in science and letters, including two Pulitzer Prizes, for *On Human Nature* (1979) and, with Bert Hölldobler, *The Ants* (1991); the U.S. National Medal of Science; the Crafoord Prize, given by the Royal Swedish Academy of Sciences for fields not covered by the Nobel Prize; Japan's International Prize for Biology; the Presidential Medal and Nonino Prize of Italy; and the Franklin Medal of the American Philosophical Society. For his contributions to conservation biology, he has received the Gold Medal of the National Audubon Society and the Gold Medal of the Worldwide Fund for Nature. Much of his personal and professional life is chronicled in the memoir *Naturalist,* which won the *Los Angeles Times* Book Award in Science in 1995. More recently, Wilson has ventured into fiction, the result being *Anthill,* published in 2010. Still active in field research, writing, and conservation work, Wilson lives with his wife, Irene, in Lexington, Massachusetts.